PRINCIPLES OF
MICROBE AND CELL
CULTIVATION

PRINCIPLES OF
MICROBE AND CELL
CULTIVATION

S. JOHN PIRT
Professor of Microbiology
Queen Elizabeth College
University of London

BLACKWELL SCIENTIFIC PUBLICATIONS
OXFORD LONDON EDINBURGH MELBOURNE

© 1975 Blackwell Scientific Publications
Osney Mead, Oxford OX2 0EL
85 Marylebone High Street, London W1M 3DE
9 Forrest Road, Edinburgh EH1 2QH
P.O. Box 9, North Balwyn, Victoria, Australia

ISBN 0 632 08150 3

First published 1975

Published in the U.S.A. by
Halsted Press, a Division of
John Wiley & Sons, Inc.
New York

Set in 10 on 12 point Monotype Times New Roman

Printed in Great Britain by
William Clowes & Sons, Limited
London, Beccles and Colchester

Contents

Preface   ix

CHAPTER 1. **Introduction**   I

Nature of a microbial culture. Historical development

CHAPTER 2. **Parameters of growth and analysis of growth
data**   4

Parameters. Growth rate. Validity of exponential growth law.
Growth yield. Metabolic quotient. Effect of substrate concentration on growth rate. Saturation constant (K_s) values. Definition of
the lag period before growth. Limits to maximum biomass concentration

CHAPTER 3. **Estimation of biomass**   15

Factors influencing choice of method. Mass and volume measurement. Mass of a cell component. Growth yields. Metabolic rates.
Light scattering. Cell and organelle counts. Staining methods

CHAPTER 4. **Batch culture and plug-flow culture**   22

Open and closed systems. Growth phases of a simple batch culture. Single point estimates of growth. Mathematical model of a
simple batch culture. Variations in the growth curve of a simple
batch culture. Plug-flow culture. Applications of plug-flow culture

CHAPTER 5. **Chemostat culture**   29

History. Theory of the chemostat. Rate of output of biomass in a
chemostat. Distribution of residence times in a chemostat. Deviations from chemostat theory. Duration of transient states after a
step change in dilution rate. Special purposes of chemostat culture

CHAPTER 6. **Elaborations on the chemostat**   42

Introduction. Turbidostat. Chemostat with feedback of biomass.
Chemostats in series. Chemostats in series with biomass feedback

CHAPTER 7. **Death of cells in growing cultures**   57

Definitions of dead and dormant cells. Death rate. Production of
dead cells at the time of division. Effects of cell death on growth

v

CHAPTER 8. Energy and carbon source requirements 63

Introduction. Assimilated carbon. Maintenance energy. Effects of maintenance energy. ATP yield (Y_{ATP}). Conditions affecting metabolic fates of carbon and energy substrates. Utilization of mixed carbon sources. Carbon dioxide supply. Carbon dioxide–carbonate equilibria in solution. Influence of carbon dioxide partial pressure on growth and metabolism. Hydrocarbons as carbon and energy sources. Dispersion of hydrocarbons in aqueous media

CHAPTER 9. Oxygen demand and supply 81

Introduction. Oxygen demand. Oxygen solubility. Measurement of amount of dissolved oxygen. Redox potential. Oxygen transfer from gas to liquid to biomass. Measurement of K_La value in absence of biomass. Measurement of K_La value during a culture

CHAPTER 10. Aeration and agitation methods 94

Introduction. Agitation and mixing. Baffled, vortex and airlift systems of aeration and agitation. Impeller design. Effect of stirrer rate. Effect of air sparging. Effects of temperature and viscosity. Effects of surface active agents and hydrocarbons. Effect of biomass. Power requirement. Foam formation and its control. Agitation-aeration systems for laboratory fermenters. Shake-flask aeration. Factors influencing oxygen solution rates in shake-flasks. Aeration in roller tubes. Aeration in static submerged cultures

CHAPTER 11. Effects of oxygen on microbial cultures 107

Introduction. Oxygen-limited growth. Effect of dissolved oxygen tension on oxygen uptake rate of growing cells. Influence of growth conditions on respiration rate of resting cells. Influence of dissolved oxygen tension on amounts of respiratory and catabolic enzymes in cells. Transitions between aerobic and anaerobic metabolism in facultative anaerobes. Effects of dissolved oxygen tension on functions other than respiration. Substitutes for oxygen. Inhibition by oxygen. Anaerobic growth.

CHAPTER 12. General nutrition 117

Introduction. Microbiological assay of growth factors. Nitrogen requirements. Vitamin and hormone requirements. Phosphorus requirements. Potassium and sodium requirements. Magnesium requirements. Sulphur requirements. Trace elements. Removal of trace elements from media. Metal ion chelation. Design of a culture medium. Conclusion

CHAPTER 13. Effects of temperature 137

Effects on growth rate. Activation energy of growth. Upper limit of growth temperature. Effects on nutrient requirements. Effects on product formation. Effects on microbial composition. Mechanisms of temperature effects

CHAPTER 14. **Effects of hydrogen ion concentration**   143

Extracellular and intracellular pH. Control of culture pH. Effects
of pH on growth and metabolism. Effects of pH on biomass composition and morphology. Molecular basis of pH effects

CHAPTER 15. **Effects of water activity and medium tonicity**   147

Introduction. Definition of water activity. Relation of water activity to solute concentration. Relation of water activity to osmotic
pressure. Measurement of tonicity and water activity. Use of
terms, tonicity and water activity. Tonicity of cell contents. Relation of growth rates to tonicity and water activity. Effects of tonicity on cell composition and metabolism. Mechanisms of tonicity
effects. Derivation of relation between osmotic pressure and
water activity

CHAPTER 16. **Product formation in microbial cultures**   156

Introduction. Relation of growth rate to product formation rate.
Product decomposition rate. Product formation in batch culture.
Product formation in a chemostat culture. Control of decay of
synthetic activity. Effects of environmental conditions on formation of microbial products

CHAPTER 17. **Effects of chemical inhibition and activation
of growth**   170

Introduction. Competitive inhibition. Non-competitive inhibition.
Product inhibition. Competitive inhibition by a product in chemostat culture. Non-competitive inhibition by a product in chemostat culture. Inhibitor affecting growth yield. Substrate inhibition
of growth. Activators of growth

CHAPTER 18. **Cultures at low and zero growth rates**   186

Stationary phase behaviour. Stationary phase of a bacterial culture. Stationary phase of a fungal culture. Minimum growth rate.
Kinetics of sporulation in *Bacillus*

CHAPTER 19. **Growth lag**   194

Introduction. Apparent lag. Causes of true lag. Inoculum effects.
Effect of temperature. Metabolic processes during lag. Conclusion

CHAPTER 20. **Mixed cultures**   199

Introduction. Competition for the same growth-limiting substrate.
Two species with different growth-limiting substrates. Product of
one species is substrate for the other. Inhibitory product of one
species is limiting substrate of the other species. Predator–prey interactions. Determination of mutation rate. Conclusion

CHAPTER 21. **Batch cultures with substrate feeds**   211

Fed batch culture. Product formation. Repeated fed batch culture.
Applications of fed batch culture. Dialysis culture. The diffusion
capsule

CHAPTER 22. **Submerged films and pellets of biomass**   223

Submerged film of biomass. Packed column with microbes
attached. Growth of submerged pellets of biomass

CHAPTER 23. **Growth of microbial colonies on the surface
of solid medium**   234

Introduction. Model of colony growth. Experimental observations on the mode of growth of bacterial colonies. Experimental
observations on fungal colony growth. Conclusion

CHAPTER 24. **Mathematical models of biomass autosynthesis**   243

Introduction. Interdependent syntheses. Automatic adjustment to
environmental change. Magnitude of an environmental effect on
growth rate. Alternative cycles of interdependent syntheses. Conclusion

CHAPTER 25. **Abstract and conclusion**   251

APPENDIX. **Symbols and abbreviations**   258

References   260

Index   269

Preface

Microbial culture in both laboratory and industry is still treated more as an art than a science, that is to say, the approach is intuitive rather than logical. The main aim of this monograph is to discuss the control of microbial cultures and to formulate as far as possible a framework of general results and principles. Such a framework is needed to unify the science and facilitate its advance and applications in brewing, food and industrial fermentation, effluent purification and elucidating the behaviour of microbes in their natural environments.

Until recently the dynamics of microbial growth were generally confined to the 'phases of growth' of a batch culture and perhaps mention of chemostat culture to control growth rate or to prolong the culture duration. Simple batch culture was the sole culture system used for the study of growth dynamics until 1950. Meanwhile much discussion and experiment has clarified the special applications and the unique advantages not only of simple batch culture, but also of chemostat culture, the chemostat with feedback of biomass, chemostats in series, plug-flow culture, the turbidostat, surface culture and, lastly, fed batch culture. Each of these systems has its place in research and industrial or community use. The present work considers mainly the developments in the last quarter of a century and analyses their significance.

The basic factors which influence microbial growth and behaviour may be classified into two groups: the intracellular factors which depend on the structure and internal milieu of the organism; and the extracellular factors or conditions external to the cell. The intracellular factors are the structure, metabolic mechanisms and genetic material of the cell and excellent treatises on these factors are to be found in cytology, biochemistry and genetics texts. In the present text the effects of extracellular factors are given the more attention. The treatment of the principles is intended to be comprehensive; however, in order to reduce the scale of the work, the following have arbitrarily been omitted: requirements of cultures for light, effects of light, and synchronization of cell division.

General principles should present the maximum amount of information in the minimum amount of time. In order to do this, for the quantitative aspects of microbial growth, mathematical models are essential. The models serve two purposes: to generalize the results, and provide a basis for the deduction of further properties of cultures. In order to keep the argument down to earth, and to reduce the mathematics to the scope of elementary algebra and calculus, not beyond the university entrance level, I have been uninhibited in making simplifying assumptions and approximations.

Elucidation of the quantitative reaction of the microbe or cell to its environment will normally progress from study of simple batch culture to chemostat culture with fed batch culture as either an intermediate or final method of culture. Although knowledge of culture behaviour has rapidly developed since 1950, there is a great lack of knowledge of the effects of temperature, pH value, oxygen and the immense variety of different nutritional conditions. Also detailed studies beyond simple batch culture are confined to a dozen or so bacterial types and even fewer of the fungi, protozoa and tissue cells out of the millions of different types in existence. Thus it seems that the scope for advance in knowledge of this field is limitless.

I have written the book for students of microbiology, in fact, most of the chapters have been developed from lectures I have given for the degree course in Queen Elizabeth College. The book is also meant to be a reference text for research workers concerned with microbial culture in the fields of fermentation, the food industry, effluent purification, waste disposal and recycling, medicine, agriculture and ecology. I have quoted only a few results of the culture of animal and plant tissue cells but this should, in the future, be a major area for the application and development of these principles.

October 1974   S. J. PIRT

CHAPTER 1

Introduction

1.1 Nature of a microbial culture

Microbe and tissue cell culture concerns the growth and functions of living matter from all three kingdoms, namely, animals, plants and protists. The term *microbes* is used as a trivial synonym for the protists which comprise the bacteria, fungi, algae and protozoa, whereas cultured tissue cells are obtained from plants and animals.

Cultures vary greatly in form and complexity. The simplest form is a homogeneous suspension of organisms of a single type in an aqueous medium containing the minimum number of nutrients, all defined, and with constant physical conditions. The complexity is added to when the culture contains more than one type of organism, multiple alternative substrates, inhibitors, the medium is heterogeneous and the conditions are transient rather than constant. As yet the theoretical principles of culture behaviour have hardly advanced beyond those for the simplest form of culture already referred to. The principles will not be established fully until we have explored the effects of all conditions over their full ranges with all types of organism not only qualitatively but also quantitatively.

1.2 Historical development

The main stream of scientific study of microbial cultivation can be traced back to the 1830's when it was discovered by Cagniard de Latour, Kützing and Schwann, that the growth of yeasts and other protists is responsible for the wine and other fermentations. This idea was for decades strongly resisted by the chemists led by Liebig. No further progress occurred until the monumental work of Pasteur started in the 1850's. He characterized the bacteria and yeasts physiologically, introduced aseptic methods, minimal media, and began to define the requirements for nutrients and oxygen.

The first completely defined medium was developed by Raulin (1869) for the growth of an *Aspergillus* species. Raulin's work was complete in that he defined the requirements for nutrients not only qualitatively but also quantitatively in terms of the growth yield; the latter aspect was, unfortunately, neglected in many subsequent nutritional studies. In the 1870's Koch

I

introduced pure culture technique to ensure that inocula contained only known species. In evaluating the remarkable work of Pasteur prior to the 1870's it should be noted that it was done without the advantage of pure culture technique. Perhaps the reproducibility of Pasteur's results depended on the selectivity of his culture media to ensure sufficient constancy of the microbial population. The work of Pasteur and his student Raulin established the requirements for the major and minor elements (Section 12.1) and for an energy source. The need for complex organic substrates now called 'growth factors' was first recognized by Wildiers (1901) through his discovery of the B vitamins or *bios* factors required for growth of yeast.

Development of more sophisticated apparatus for culture growth began in the 1930's with the introduction of the shake-flask technique by Kluyver & Perquin (1933). This advance provided a much more convenient laboratory method for aerated submerged culture and the first means for submerged culture of aerobic fungi, which previously had been grown only on the surface of a solid or liquid medium. In surface culture the environment of the organism is heterogeneous, whereas in submerged culture it is homogeneous and therefore simpler to understand and control. Subsequently the submerged culture of aerobic fungi proved essential in the development of the antibiotics industry. The growing technological importance of microbial culture in the 1940's and 1950's stimulated much engineering interest which led to the development of the stirred fermenter with automatic controls of the culture environment. Such equipment is now central to the elucidation of microbe and tissue cell behaviour.

Little development of culture kinetics occurred before 1950. Slator (1921) originated concepts of the interrelations of culture parameters. Hinshelwood (1946) contributed useful models of the kinetics of living cell reactions. The monograph of Monod (1942) is a landmark in that it showed how bacterial growth could be formulated in terms of the growth yield, the specific growth rate and the concentration of growth-limiting substrate. From this analytical approach the theory of the chemostat type of continuous-flow culture followed (Monod, 1950; Novick & Szilard, 1950). Chemostat culture is now an essential means for elucidating the relations between an organism and its environment.

Animal cell culture has made rapid progress since about 1950 particularly through the development of monolayer culture, trypsin treatment to disperse cells and suspended cell culture, and the elucidation of the nutrient requirements of cells. Plant cell culture has attracted less attention and its quantitative basis seems less sure than that of animal cell culture.

The development of microbial culture theory has been remarkably slow compared with that of the theory of physics or chemistry or economics, each of which, even before 1900, had been expanded by numerous great individual workers into a large corpus of knowledge. This cannot be because of a lesser need for an understanding of the response of the cell to its environment, both in the short and long term. Since 1950 the study of open culture systems, of which the chemostat is one of the most important, has revealed new horizons which give vast scope for both theoretical development and experimental testing of theory.

CHAPTER 2

Parameters of growth and analysis of growth data

2.1 Parameters

The qualitative observation of whether or not growth occurs in a culture is useful for many purposes. The further information obtained when the growth is measured quantitatively is often presented in the form of a graph of biomass against time. The data may be rendered more meaningful and concise if it is analysed in terms of the various growth parameters: specific growth rate or doubling time of the biomass, growth lag, growth yield, metabolic quotients for substrate utilization and product formation, substrate affinity and maximum biomass.

The growth parameters are defined with reference to the growth of a simple homogeneous batch culture. Such a system is supposed to consist of a well-mixed batch of inoculated medium. It is assumed that the mixing is adequate to disperse the biomass and that the medium is free from concentration gradients. A heterogeneous culture such as a colony growing on a surface clearly is more complex (Chapter 23).

2.2 Growth rate

2.2.1 Specific growth rate

The requisite conditions for growth of biomass in a culture are: (i) a viable inoculum, (ii) an energy source, (iii) nutrients to provide the essential materials from which the biomass is synthesized, (iv) absence of inhibitors which prevent growth, (v) suitable physicochemical conditions.

If all the requirements for growth are satisfied, then during an infinitely small time interval (dt) one expects the increase in biomass (dx) to be proportional to the amount (x) present and to the time interval; that is

$$dx = \mu x.dt \qquad\qquad 2.1$$

hence

$$dx/dt = \mu x \qquad\qquad 2.2$$

The differential coefficient (dx/dt) expresses the population growth rate. The parameter μ which represents the rate of growth per unit amount of biomass $(1/x)(dx/dt)$, is termed the *specific growth rate* and has the dimension of reciprocal time $(1/t)$. It is analogous to the compound interest rate on an investment—thus a specific growth rate of $0 \cdot 1 \ h^{-1}$ is equivalent to a compound interest rate of 10% per hour.

When μ is constant, integration of Eqn 2.2 gives

$$\ln x = \ln x_0 + \mu t \qquad\qquad 2.3$$

where x_0 is the biomass when $t = 0$. The plot of $\ln x$ against time will be a straight line with slope, μ. If the logarithms are converted to the base 10 then Eqn 2.3 becomes

$$\log x = \frac{\mu t}{2 \cdot 30} + \log x_0 \qquad\qquad 2.4$$

By putting Eqn 2.3 in the form

$$\ln (x/x_0) = \mu t \qquad\qquad 2.5$$

it follows that

$$x = x_0 \, e^{\mu t} \qquad\qquad 2.6$$

Growth which accords with this law is called constant exponential or logarithmic growth. The basic measure of the growth rate is the specific growth rate. The other measures which are given below can be related to the specific growth rate.

2.2.2 Biomass doubling time

The relation between specific growth rate and the doubling time (t_d) of the biomass is obtained from Eqn 2.5 by putting $x = 2x_0$ and $t = t_d$, hence

$$t_d = \frac{\ln 2}{\mu} = \frac{0 \cdot 693}{\mu} \qquad\qquad 2.7$$

2.2.3 Degree of multiplication

The degree of multiplication is given by x/x_0 and is equal to $e^{\mu t}$ (see Eqn 2.6). Alternatively, if the biomass undergoes n doublings or generations, we can write

$$x/x_0 = 2^n \qquad\qquad 2.8$$

therefore

$$n = 3 \cdot 32 \log (x/x_0) \qquad\qquad 2.9$$

Often in culture experiments it is useful to use an inoculum size which is 10% of the final biomass, n will then be $3 \cdot 32$.

2.2.4. Reciprocal doubling time

Since the number of doublings of the biomass in time, t, will be t/t_d we can write

$$x = x_0 2^{t/t_d} \qquad\qquad 2.10$$

Taking logarithms to the base 2, Eqn 2.10 becomes

$$\log_2 x = \log_2 x_0 + t/t_d \qquad\qquad 2.11$$

the slope of which is $1/t_d$, that is the reciprocal doubling time. Some workers have preferred to express growth rates in terms of $1/t_d$ rather than specific growth rate. However, μ seems preferable since it occurs in the fundamental growth law (Eqn 2.2), and the convenience of this convention is shown by its universal adoption in the theory of continuous flow culture. It is important to make clear which convention is being followed.

2.3 Validity of exponential growth law

The law of constant exponential growth must be followed when the environmental conditions are constant and the constitution of the biomass remains constant. Conversely, if the growth does not accord with the law, it follows that either one or both of the two qualifying conditions are not being met. Bacterial and yeast cultures are well known to have exponential growth phases but the law is universal for all procaryotes and eucaryotes provided that the above stated conditions are met. The law has been demonstrated to apply to protozoan cultures (Phelps, 1936) and to animal cells in culture (Birch & Pirt, 1970). The applicability of the exponential growth law to mould cultures in homogeneous submerged culture has been established (Pirt & Callow, 1960). Growth of filamentous fungi can easily show divergence from the exponential either because they form pellets in which the biomass is not homogeneously dispersed in the medium or because oxygen supply becomes growth-limiting. Probably the most common first cause of failure to maintain constant exponential growth is a change in the environment. This is inevitable sooner or later in a batch culture. Further it is probable that any change in the environment will subsequently induce a change in the constitution of the biomass. A simple model of growing biomass indicates that, in response to a changed environment, the proportions of the various components of the biomass will automatically be adjusted so as to maximize the growth rate (Chapter 24).

2.4 Growth yield

The growth yield is defined by the quotient

$$\Delta x/\Delta s = Y \qquad\qquad 2.12$$

where Δx is the increase in biomass consequent on utilization of the amount Δs of substrate. More rigorously the growth yield is expressed by the limit of $\Delta x/\Delta s$ as $\Delta s \rightarrow 0$, that is

$$Y = dx/ds \qquad\qquad 2.13$$

It should be noticed that if x and s are the biomass and substrate concentrations respectively, then $Y = -dx/ds$, the negative sign being introduced because x and s vary in opposite senses.

The growth yield is important as a means of expressing the quantitative nutrient requirement of an organism. As such the growth yield was used by Raulin (1869) to express the nutrient requirements of a fungus; subsequent workers in mycology have tended to use the reciprocal of Y called the *economic coefficient*. Monod (1942) first showed that in bacterial cultures, when the conditions are maintained constant, the growth yield is a constant, reproducible quantity. Thus if x_0 and s_0 are the initial biomass and substrate concentrations respectively and x and s are the corresponding concentrations during the growth of the culture

$$x - x_0 = Y(s_0 - s) \qquad\qquad 2.14$$

For a growth-limiting substrate, when the culture has reached its maximum biomass (x_m), $s \approx 0$ and we may write

$$x_m - x_0 = Y s_0 \qquad\qquad 2.15$$

Hence, for a growth-limiting substrate, a plot of x_m against s_0 should be a straight line with slope, Y. Figure 2.1 is an example of the application of this relation to determine the growth yield of cultured animal (mouse) cells for choline.

2.5 Metabolic quotient

The rate of consumption of a substrate in a culture at a particular moment is given by

$$ds/dt = qx \qquad\qquad 2.16$$

where x is the biomass and the coefficient q is known as a metabolic quotient, or specific metabolic rate. Well-known examples are q_{O_2} and $q_{glucose}$ for oxygen and glucose consumption respectively. The metabolic quotient is analogous to an enzyme activity. If the biomass constitution is constant and the environment is constant then q must be constant.

We can write for the substrate consumed for growth in the small time interval dt

$$ds = \frac{\mu x}{Y} \cdot dt \qquad\qquad 2.17$$

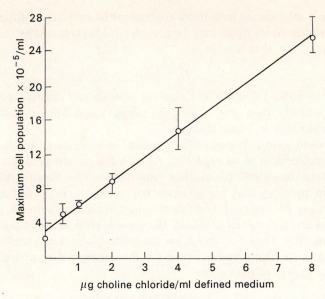

FIG. 2.1 The effect of choline concentration on the growth of mouse LS cells in culture. The points on the graph are the mean values; vertical lines represent the ranges of the results. (From Birch & Pirt, 1969)

hence

$$ds/dt = \mu x / Y \qquad\qquad 2.18$$

and comparison of Eqn *2.18* and *2.16* shows that

$$q = \mu / Y \qquad\qquad 2.19$$

Equation *2.19* is used to estimate the demands for substrates (especially oxygen) at different growth rates.

The use of metabolic quotients to express rates of product formation is considered in Chapter 16.

2.6 Effect of substrate concentration on growth rate

The occurrence of constant exponential growth in batch cultures shows that growth rate may be virtually unaffected by substrate concentration over a wide range; that is the growth process shows zero order kinetics. We might expect that substrate consumption would follow enzyme kinetics so that if s is the substrate concentration and q is the metabolic quotient

$$q = q_m s / (s + K_s) \qquad\qquad 2.20$$

where K_s, called the *saturation constant*, is equivalent to a Michaelis–Menten constant and q_m is the maximum value of q obtained when $s \gg K_s$. If we make the substitutions, $q = \mu/Y$ and $q_m = \mu_m/Y$, then it follows that

$$\mu = \mu_m s/(s + K_s) \qquad\qquad 2.21$$

Equation *2.21* is often termed the Monod relation since Monod (1942) first showed empirically that the expression accorded well with the relation of bacterial growth rate to substrate concentration (Fig. 2.2). The K_s value is in-

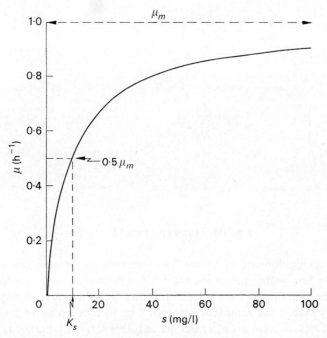

FIG. 2.2   Specific growth rate (μ) plotted as a function of substrate concentration (s) according to the Monod equation $\mu = \mu_m s/(s + K_s)$ where $\mu_m = 1 \cdot 0 \ \mathrm{h^{-1}}$ and $K_s = 10 \ \mathrm{mg/l}$. Note, When $s = K_s$, $\mu = 0 \cdot 5 \mu_m$.

versely related to the affinity of the organism for the substrate. Rearranging Eqn *2.21* we obtain

$$\frac{1}{\mu} = \frac{K_s}{s\mu_m} + \frac{1}{\mu_m} \qquad\qquad 2.22$$

hence a plot of $1/\mu$ against $1/s$ should give a straight line with an intercept on the abscissa at $-1/K_s$, and an intercept on the ordinate at $1/\mu_m$ (Fig. 2.3). The linearity of the double reciprocal plot forms a convenient test of the validity of the hyperbolic relation.

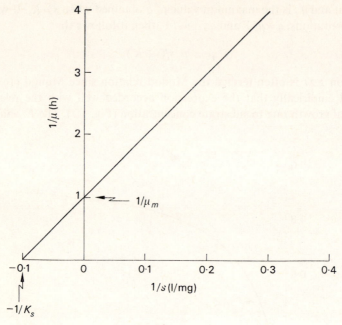

FIG. 2.3   Reciprocal plot of data for Fig. 2.2.

2.7 Saturation constant (K_s) values

Few determinations of K_s values have been made, because the values are extremely low—often below the sensitivity of chemical assay methods—and unless the sampling or assay is instantaneous the substrate level may fall substantially before the sample is assayed. The methods used to measure the values of s at different growth rates are as follows: (*i*) instantaneous measurement of the growth-limiting substrate in a chemostat culture at different growth rates as for oxygen (Johnson, 1967b); (*ii*) measurement of initial growth rates with different substrate concentrations in batch cultures; (*iii*) measurement of the growth rates at the end of a batch culture (Monod, 1942) and estimation of the substrate concentration from the biomass by means of Eqn 2.14; (*iv*) measurement of the critical dilution rates in chemostat cultures with different concentrations of the growth-limiting substrate in the feed medium (Dean & Rogers, 1967); (*v*) the steady state method of Button (1969), illustrated in Fig. 2.4.

Table 2.1 shows that the K_s values for carbon and energy sources are generally about 10^{-5} M. The values for phosphate, potassium and magnesium ions also are about 10^{-5} M and the value for oxygen is 10^{-6} to 10^{-5} M. The values for amino acids and vitamins are lower by one or more orders of

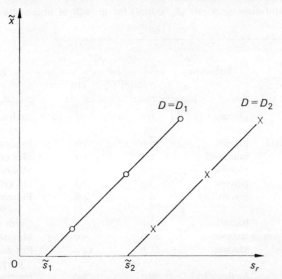

FIG. 2.4   Method of Button (1969) to determine the concentration of growth-limiting substrate ($\tilde{s}$) at a given specific growth rate ($\mu = D$) in a chemostat. Each straight line is a plot of the steady-state biomass ($\tilde{x}$) at a given dilution rate with different concentrations (s_r) of growth-limiting substrate in the feed medium. The straight line is represented by $\tilde{x} = Y(s_r - \tilde{s})$ where Y is the growth yield. On extrapolation of the line to $\tilde{x} = 0$ we have $s_r = \tilde{s}$.

magnitude. No K_s values for the trace elements are available and the validity of the Monod relation for trace element effects on growth rate has not been established. The K_s values are of the same order for both procaryotes and eucaryotes. For the high polymer starch it appears that, on a molar basis, the K_s value for $K.$ $aerogenes$ is about the same as for the monomer (Hernandez & Pirt, 1975). For the bacterial substrate of the protozoan $Tetrahymena$ (Table 2.1) the K_s value on a weight basis is comparable to that of soluble small molecules.

In general, the results follow the Monod relation (Eqn 2.21). However, Shehata & Marr (1971) found that deviations from the Monod relation occur at the higher growth rates which, they suggested, indicates that the organism's affinity for the growth-limiting substrate varies with the growth rate.

2.8 Definition of the lag period before growth

There is often a lag before the specific growth rate of the biomass reaches its maximum value. The duration of this lag (L) is, for most purposes, conveniently defined by the method of Lodge & Hinshelwood (1943). In the plot of the

TABLE 2.1   Saturation constants (K_s values) for growth of organisms on diverse substrates

Organism (genus)	Substrate	K_s (mg/l)	K_s (10^{-5} M)	Reference
Escherichia	glucose	6.8×10^{-2}	3.8×10^{-2}	Shehata & Marr (1971)†
Unknown bacteria	phenol	9.0×10^{-1}	1.0	Jones et al. (1973)
Escherichia	mannitol	2.0	1.1	Monod (1942)
Escherichia	glucose	4.0	2.2	Monod (1942)
Aspergillus	glucose	5.0	2.8	Pirt (1973)
Candida	glycerol	4.5	4.9	Button & Garver (1966)
Tetrahymena	bacteria	12.0	—	Curds (1971)
Escherichia	lactose	20.0	5.9	Monod (1942)
Saccharomyces	glucose	25.0	14.0	Pirt & Kurowski (1970)
Pseudomonas	methanol	0.7	2.0	Harrison (1973)
Pseudomonas	methane	0.4	2.6	Harrison (1973)
Klebsiella	carbon dioxide	0.4	9×10^{-1}	Section 8.8
Escherichia	phosphate ions	1.6	1.7	Shehata & Marr (1971)†
Klebsiella	magnesium ions	5.6×10^{-1}	2.3	Dean & Rogers (1967)
Klebsiella	potassium ions	3.9×10^{-1}	1.0	Dean & Rogers (1967)
Klebsiella	sulphate ions	2.7	2.8×10^{-1}	Dean & Rogers (1967)
Candida	oxygen	4.5×10^{-1}	1.4	Button & Garver (1966)
Candida	oxygen	4.2×10^{-2}	1.3×10^{-1}	Johnson (1967)
Aspergillus*	arginine	5.0×10^{-1}	2.9×10^{-1}	Pirt (1973)
Escherichia*	tryptophan	1.1×10^{-3}	5.4×10^{-3}	Novick (1958)
Escherichia*	tryptophan	4.9×10^{-4}	3.4×10^{-3}	Shehata & Marr (1971)
Cryptococcus	thiamine	1.4×10^{-7}	4.7×10^{-10}	Button (1969)

* Amino acid auxotroph
† The lower of the two values found

biomass logarithm (Fig. 2.5) the straight line is extrapolated to the initial biomass level and the intercept on the time axis is taken to be L. Hence the growth of the culture is represented by

$$\ln x = \mu(t-L) + \ln x_0 \qquad\qquad 2.23$$

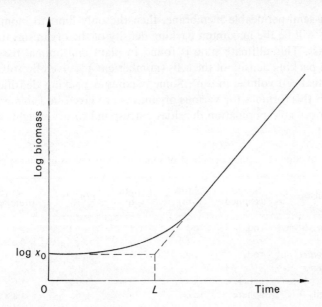

FIG. 2.5 Definition of lag period, L.

2.9 Limits to maximum biomass concentration

The maximum density of biomass of any kind which can be reached in a given medium is determined by one of the following four conditions: (i) the amount of growth-limiting substrate supplied, (ii) the accumulation of an inhibitory product, (iii) the maximum packing density of the biomass, (iv) cell death. Bail (1929) proposed that the maximum bacterial density or 'M' concentration in a culture was determined by the requirement for *lebensraum* or biological space. This vague concept implied that the biological space exceeded the physical space occupied by the organisms. However, it is clear that the M concentrations suggested by Bail can be accounted for by either nutrient (especially oxygen) limitation or an inhibitory product in the cultures. Bail's idea is of historical interest since it held sway for a number of years among bacteriologists and we can see a parallel to it in the concept of 'contact inhibition' of growth in animal tissue cell cultures. Cells said to be subject to contact inhibition have been found to grow indefinitely in multilayer form when supplied continuously with nutrients (Kruse & Miedema, 1965), also other cells said to have growth stopped by contact, have been found to be inhibited by acid production (Ceccarina & Eagle, 1971); thus the concept that contact between animal cells prevents their growth has been disproved.

Provided that there is no inhibition, the population remains viable, and there is an unlimited supply of nutrients, which might be arranged by diffusion through a semi-permeable membrane, then the only limit to biomass concentration will be the maximum packing density of the organisms or cells in the biomass. This ultimate state is found in plant and animal tissues. The maximum packing density of the cells (number/cm^3) is given by $10^{12}/v$ where v=individual cell volume in μm^3. Some maximum packing densities calculated from this formula for various organisms are given in Table 2.2. These maximum possible population densities correspond to dry weights of about 10% (w/v).

TABLE 2.2   Maximum packing densities* of some microorganisms in number per cm^3

Organism	Shape assumed	Radius (μm)	Length (μm	$v(\mu$m$^3)$	Maximum (number/cm^3)
Serratia marcescens†	rod	0·5	1·7	1·34	$7·5 \times 10^{11}$
Klebsiella aerogenes†	rod	0·86	5·4	12·5	$8·0 \times 10^{10}$
Bacillus megaterium†	rod	1·2	7·6	34·4	$2·9 \times 10^{10}$
Saccharomyces cerevisiae	sphere	3·5	—	179	$5·6 \times 10^{9}$
Mouse L cells	sphere	10·0	—	4190	$2·4 \times 10^{8}$

* Calculated from the formula, $10^{12}/v$, where v=cell volume in μm^3
† Dimensions from Powell (1963)

CHAPTER 3

Estimation of biomass

3.1 Factors influencing choice of method

Biomass is a general term used to refer to the organism in culture. The alternative terms, organisms, cells, mycelium and plasmodium, imply special kinds of organization of the biomass. The biomass parameter measures the extent of growth and it enters into certain important derived parameters such as growth yields and metabolic quotients. Despite the fundamental importance of biomass, published studies on microbial cultivation often have not included an explicit statement of the amount of biomass formed, which seriously limits the quantitative interpretation of the results.

The methods used to measure biomass are based on eight types of measurements: mass, volume or linear extent, mass of a biomass component, mass of substrate consumed or product formed, metabolic rates, light scattering, cell and organelle counts, staining methods.

Choice of the method of measuring biomass is a crucial decision to make in the approach to a culture problem and often the limitations of the methods make the decision difficult. The factors which influence the choice are: (*i*) the properties of the biomass, (*ii*) the properties of the culture medium, (*iii*) the accuracy required, (*iv*) the sensitivity required, (*v*) the required speed of measurement.

Properties of the biomass which affect the choice are: whether it is filamentous or particulate, how easily it can be separated from the medium, and the age of the biomass or its growth rate. Separation of bacteria from stable suspensions by centrifugation can be tedious; it may be possible in the future to facilitate this step by means of chemical treatment to precipitate or flocculate the bacteria.

Properties of the medium which may affect the method of biomass assay include the viscosity, colour, presence of solids or dissolved matter which reacts like biomass in the assay, presence of storage products such as glycogen and poly-β-hydroxybutyrate in the biomass. The methods vary greatly in their sensitivity, speed and accuracy. A comparison of the sensitivities of the more commonly used methods for bacterial biomass measurement is given in Table 3.1. The least sensitive method is dry weight measurement and the most sensitive is a cell count.

Despite the many methods of biomass measurement available, new and improved methods need to be sought to enable biomass to be determined under all the different conditions that can occur.

TABLE 3.1   Comparison of the sensitivities of some methods of bacterial biomass estimation

Method	Minimum dry mass of bacteria required for an estimation with an error of $< 2\%$ (mg)
Dry weight	50
Biuret protein	1·0
DNA	1·0
Folin–Ciocalteu protein	10^{-1}
Opacity	10^{-1}
Cell count	10^{-5}

3.2 Mass and volume measurement

3.2.1 Dry mass

Direct estimation of the dry biomass involves separating the organism from the medium, washing the organism and drying it. Washing of the biomass must be carried out in such a way as to prevent lysis of the organisms through rupture by osmotic shock. This may occur if the biomass is washed with water especially when the organisms are taken from a rapidly growing culture. The precaution against lysis is to wash with a near isotonic saline and allow for the dry weight of the salt present after drying. Usually biomass is dried in an oven at $105 \pm 2°C$.

The method will be inapplicable if the medium contains an indeterminate amount of other solids besides the biomass. Sometimes it may be necessary to correct the dry weight of biomass for its content of storage product which may account for up to 50% of the dry weight. The dry weight is probably the most unequivocal parameter of biomass. Its chief disadvantages are that the method requires a rather large sample and it is slow.

3.2.2 Wet mass

If, in the determination of biomass by direct weighing, the final drying operation is omitted, the wet mass is obtained. The wet mass will include both intracellular and extracellular or interstitial water. Roberts *et al.* (1955, p. 63) estimated the interstitial volume of close-packed cells of *Escherichia coli* to be 10 to 20% of the total volume, and they estimated (*ibid.*, p. 5) the dry biomass to be 25% of the wet mass. The centrifugation or other method used to pack down the biomass should be carefully standardized. The wet mass method is not as accurate as the dry mass method but it is quicker.

3.2.3 Volume

The amounts of biomasses can be compared by means of volume measurements since the density varies little. Small volumes of cells can be measured by the haematocrit method which Waymouth (1956) applied to follow the growth of animal cells in culture. In the case of voluminous biomasses, such as may be obtained in fungal cultures, volume measurement in a graduated centrifuge tube is often used as a quick method. It may be advantageous when the medium contains other solids such as calcium carbonate which has a high density compared with the biomass. As in wet mass determination, the method of packing down the biomass needs to be standardized.

3.2.4 Linear extent

The growth of colonies or pellets of biomass can be followed by measurement of the linear spread of the colony or pellet. The calculation of growth rates from the data is discussed in Chapter 23.

3.3 Mass of a cell component

The amount of a cell component, which is a constant proportion of the total biomass, may be used as a measure of the biomass. Components which may be used for this purpose are cell nitrogen, protein and DNA. The RNA content of the cell may be a useful biomass parameter provided that the growth rate and temperature are constant.

3.3.1 Cell nitrogen

The nitrogen content of the biomass can be estimated to within about 1% by the Kjeldahl method. Proportionality between cell nitrogen and dry biomass may only occur under a limited range of conditions, e.g. the nitrogen content of dry mycelium of *Penicillium chrysogenum* can vary from 10% in the fast growing organism to 6% in stationary phase mycelium.

3.3.2 Protein

Two methods are commonly used for the determination of biomass protein after it has been solubilized by alkali: one is based on the biuret reaction of peptide bonds (Stickland, 1951; Layne, 1957, p. 447); the other is based on the Folin Ciocalteu reaction of tyrosine and tryptophan residues (Lowry *et al.*, 1951; Layne, 1957, p. 447). A third method based on binding of the dye bromsulphalein by basic groups on the protein shows promise (Paul, 1965, p. 331). The reagents are generally standardized against bovine serum albumin. The assay value depends upon the amino acid composition of the protein, however, the assumption that this is invariable may not always hold. The Folin Ciocalteu method is more sensitive than the biuret method (see Table 3.1). Khanna *et al.* (1963) found the Lowry method best for the determination of fungal protein.

3.3.3 DNA

The estimation of DNA in biomass is based on the colorimetric estimation of deoxyribose. Application of this method to bacterial mass determination is described by Meynell & Meynell (1965, p. 5) and to tissue cell cultures by Paul (1965, p. 330). The DNA content of biomass, because of its constancy and specificity, should be a good measure of the biomass, but it has been little used possibly because it is slower and less sensitive than protein assays.

3.4 Growth yields

The amount of biomass formed (Δx) may be estimated from the amount of a substrate utilized (Δs) or the amount of a product formed (Δp). Ideally direct proportionality is to be preferred, that is $\Delta x = Y_{x/s} \Delta s$ or $\Delta x = Y_{x/p} \Delta p$ where the growth yields, $Y_{x/s}$ and $Y_{x/p}$, are constant, but this may not be true if the culture conditions vary. End products of energy metabolism such as lactic acid or carbon dioxide may be suitable products to relate to biomass production. Sources of carbon, energy, nitrogen or growth factors may be suitable substrates. Often the growth yields are growth rate dependent, however, this effect may be eliminated in chemostat cultures where the growth rate is maintained constant.

3.5 Metabolic rates

The amount of biomass (x) may be related to some metabolic rate. If q is the metabolic quotient and the metabolic rate is dy/dt where y is the amount of some substrate or product then $x = (1/q)(dy/dt)$. Examples of processes which may be used in this way are dye reductions and gas production. The rate of reduction of tetrazolium to form the red formazan has been used to measure the amount of animal cells in a culture (Tjötta, 1966).

3.6 Light scattering

3.6.1 History

The introduction of bacterial mass determination based on light scattering is of historic importance since it made possible for the first time instantaneous measurement of biomass, which has proved invaluable in the elucidation of the nature of cell growth. Suspensions of organisms scatter light and appear turbid when the refractive index of the organisms differs from that of the medium. Visible turbidity begins to develop when a bacterial density reaches about 10^6/ml. The turbidity may be estimated either by measurement of the light transmitted (absorptiometry) or of the light scattered (nephelometry). The method has been applied to suspensions of bacteria, yeasts, spores and mammalian cells.

3.6.2 Factors which influence amount of light scattered

A number of important physical and biological factors, not all of which are generally recognized, can affect the magnitude of the turbidity of a suspension of organisms. The organism concentration (x) and length of the light path (l) are related to the intensities of incident light (I_o) and transmitted light (I_t) by the expression $\log (I_o/I_t) = Axl$. The value of $\log (I_o/I_t)$ is termed the *opacity*, *optical density* or *extinction*. The factor A is constant at low bacterial concentrations but begins to decrease at higher concentrations because of secondary scattering; that is, light encountering more than one particle. In nephelometry if I_s = intensity of scattered light, $\log (I_o/I_s) = -Bxl$ where B is a constant. Absorptiometry is used more commonly than nephelometry probably because of the wider availability of absorptiometers.

The degree and direction of light scattering depend upon the size and shape of the particles, the wavelength of the light and the difference between the refractive index of the particles and that of the medium. These effects have been expressed by Powell (1963) in two rules.

1.   The total scattered light increases with increase in the ratio of particle size to wavelength. For this reason the lowest wavelength of light should be selected. In practice, green light of wavelength about 540 nm is commonly used; at lower wavelengths the absorption of light may be excessive.

2.   The total scattered light is greater, the greater the difference between the refractive index of the particles and that of the medium. The effect of refractive index of the organism is shown by the difference in contrast between bacterial colonies and the medium on an agar plate. Staphylococci are more contrasting than *Escherichia coli* because the former have the higher refractive index. The concentration of solutes in the medium can considerably affect its refractive index. Means of correcting the opacity for change in the refractive

index of the medium have been devised by Gilby & Few (1959). Osmotic swelling of bacteria in a hypotonic medium can considerably decrease the opacity, probably through lowering the refractive index of the cells (Mager *et al.*, 1956; Gilby & Few, 1959). The swelling is indicative of the elasticity of some cell walls, for instance *Escherichia coli* shows this effect but not staphylococci (Mitchell & Moyle, 1956). To avoid osmotic and refractive index effects on opacity the tonicity of the diluting fluid should be about the same as that of the suspension of cells.

There have been claims that the opacity $\propto$ cell concentration irrespective of the cell number, however, no generalization can be made on this point, though in special cases it may be true (Powell, 1963).

Rosenberger & Elsden (1960) found that the ratio of opacity—dry biomass for *Streptococcus faecalis* was unaffected by the specific growth rate, which might have been expected to affect the ratio because of the increase in cell size with growth rate. In batch cultures of *Klebsiella aerogenes* the ratio, opacity–bacterial dry weight was found to fall by 8% after the culture entered the stationary phase (Hadjipetrou *et al.*, 1964). Sometimes it is necessary to add a germicidal compound to the cell suspension before measuring the opacity. Formalin (1% v/v) is frequently used for this purpose. The formalin treatment was found to prevent the response of opacity to change in the tonicity of the medium (Mager *et al.*, 1956), hence this treatment may be a useful means of stabilizing the opacity.

Manfredini & Wang (1972) have reported a means for the application of turbidity measurement to determine the biomass in a hydrocarbon emulsion: the oil–water emulsion is clarified by the addition of propionic acid.

3.7 Cell and organelle counts

3.7.1 General features and total counts

Biomass can be estimated either by counting the total number of individual organisms or organelles such as nuclei present in the sample by means of a microscope or some electronic device, or by a viable count which depends on growth of the individuals into colonies. A disadvantage of the counting method is that there is a large unavoidable sampling error if the number in the sample is small. The advantages of the method are its specificity and the fact that it is the most sensitive method (Table 3.1).

The replica counts conform to a Poisson distribution. If n is the number of organisms counted, the standard deviation is given by $\sigma = n^{1/2}$ and the 95% confidence limits are taken to be $n \pm 2\sigma$. This means that n must have a value of more than 400 if the 95% confidence limits are to be less than $\pm 10\%$ of the mean. For a fuller account of the errors in counts reference should be made to Meynell & Meynell (1965).

Visual counts have the advantage that the observer can often distinguish between the species of organism to be counted and other types of particle or organism in the medium. Animal cell numbers have been estimated by counts of nuclei released from the cells by 0·1 M citric acid and stained with crystal violet (Sanford *et al.*, 1951; Paul, 1965, p. 327).

The electronic counting devices such as the Coulter counter are convenient for large numbers of counts and decrease the sampling error since much larger numbers can be counted. With cells as small as bacteria careful precautions must be taken to avoid interference by the background 'noise' (Hobson & Mann, 1970).

3.7.2 Viable cell counts

The basic methods used to count microbes by allowing the individual organisms to multiply and produce colonies which can be counted are described by Meynell & Meynell (1965). Rich rather than minimal media should be used for viable counts because isolated individual cells often are more exacting nutritionally than a dense population. By means of selective media the method can be used to enumerate organisms of different types in a mixture.

The dilution count method, familiar in the bacteriological testing of water supplies, consists of estimating the proportion of tubes of liquid medium which remain devoid of growth after inoculation from the organism suspension. The advantage of the method is that it can estimate small concentrations of organisms (< 1/ml). However, because the sample size is small the error is large. The method of collecting the organisms on a membrane filter and letting them grow into visible colonies on the filter is superseding the dilution count method.

3.8 Staining methods

The most unequivocal method of estimating the numbers of viable cells and dead cells is by comparing a colony count and a total count. In some cases the dead and viable cells can be differentiated by staining with a dye. Viable cultured mammalian cells are impermeable to trypan blue whereas the dead cells take up the dye. Eosin is used similarly in yeast cell cultures.

Stains which are only taken up by viable cells are called vital stains. Staining mammalian cells in culture with neutral red is used as a quantitative means to estimate the cells surviving viral attack (Finter, 1969).

CHAPTER 4

Batch culture and plug-flow culture

4.1 Open and closed systems

Microbial cultures can be classified into 'open' and 'closed' systems. In an open system all the materials which compose the system may enter and leave it. The system is closed if some essential part of the system cannot both enter and leave it. Continuous-flow cultures, therefore, which have input of nutrient medium and output of biomass and other products are open systems. A simple batch culture which consists of an initial limited amount of nutrient medium is an example of a closed system. In a closed system the growth rate of the biomass must tend towards zero, either because of lack of a nutrient or because the further accumulation of a product can be no longer tolerated; hence such systems are always in transient states. In contrast, in the open system there is the possibility that the rate of conversion of substrate to products and biomass will balance the output rate, that is a steady state may be established.

4.2 Growth phases of a simple batch culture

In a simple homogeneous batch culture all parts of the culture are subject to the same conditions. The different phases which may occur in such a culture are depicted in Fig. 4.1. These phases reflect changes in the biomass and in its environment. After the lag period (Chapter 19) growth occurs at the maximum rate and finally ceases either through lack of a nutrient, or accumulation of an inhibitory product or some change in the physical environment. After the biomass reaches its maximum, there may be a stationary phase where the amount of biomass remains constant but, sooner or later, the biomass declines in amount as a result of maintenance metabolism or autolysis.

The duration of the exponential growth phase depends partly on the initial concentration of growth-limiting substrate. If this were 1 g/l and $K_s = 4$ mg/l it follows from Eqn 2.21 that $\mu > 0.95\,\mu_m$ until 92% of the growth-limiting substrate is consumed. Consequently the period when nutrient limitation significantly affects the growth rate will be limited to a small fraction of the last generation. This behaviour is typical of most simple batch cultures and means that they are subject to an extreme type of nutritional regimen, that is growth with excess substrate followed by sudden starvation. Consequently the

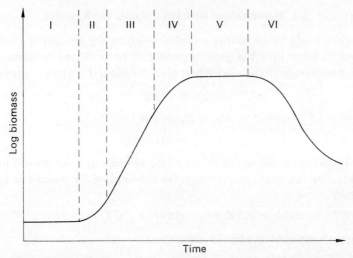

FIG. 4.1 Batch growth curve with six phases: I, lag; II, accelerating growth; III, exponential growth; IV, decelerating growth; V, stationary; VI, decline.

periods of growth at rates less than the maximum are not long enough to permit the organism to adjust its structure to that which is optimal for the growth rate. This limitation can, to some extent, be overcome by elaborations on batch culture (Chapter 21), but it can only be overcome fully by means of chemostat culture.

The behaviour of non-growing biomass, which is present in the stationary and decline phases, is described in Chapter 18. Apparent decelerating growth, stationary or decline phases may occur if there is onset of some lethal condition, so that the net growth rate is the difference between the rates of growth and death of cells (Chapter 7).

4.3 Single point estimates of growth

Sometimes the extents of growth in batch cultures are compared, not by means of growth curves but by single point estimates of growth after an arbitrary time. It should be noted that differences in single point estimates can be interpreted in several ways since they may reflect differences in lag periods, maximum specific growth rates, maximum population densities, concentrations of inhibitory products or biomass decline rates. Hence, unless it is known from other evidence, that certain of these interpretations can be excluded, single point estimates of growth must be regarded as combinations of all these effects, and they do not allow particular growth parameters to be evaluated.

4.4 Mathematical model of a simple batch culture

When growth of a batch culture is limited solely by the amount of a substrate supplied initially, then the growth curve can be predicted in terms of the growth parameters. In the model, first given by Monod (1942, p. 123) we have

$$dx/dt = \mu x \qquad 4.1$$

$$\mu = \mu_m s/(s+K_s) \qquad 4.2$$

$$x-x_0 = Y(s_0-s) \qquad 4.3$$

where x_0 and s_0 are the initial values of the biomass and the growth-limiting substrate concentrations respectively. On substituting for μ and s in Eqn 4.1 we obtain

$$dx/dt = \mu_m(Y_s+x_0-x)x/(K_s Y+s_0 Y+x_0-x) \qquad 4.4$$

We obtain x in terms of t by the integration

$$\int_{x_0}^{x} \frac{(K_s Y+s_0 Y+x_0-x)\,dx}{Ys_0+x_0-x} = \mu_m \int_0^t dt \qquad 4.5$$

By conversion to partial fractions, the left-hand side of Eqn 4.5 becomes

$$P\int_{x_0}^{x} \frac{dx}{x} + Q\int_{x_0}^{x} \frac{dx}{Ys_0+x_0-x} \qquad 4.6$$

where $P=(K_s Y+s_0 Y+x_0)/(Ys_0+x_0)$ and $Q=K_s Y/(Ys_0+x_0)$. The solution of Eqn 4.5 is

$$P\ln(x/x_0) - Q\ln\{(Ys_0+x_0-x)/Ys_0\} = \mu_m t \qquad 4.7$$

Equation 4.7 gives the familiar s-shaped curve of batch growth in which the value of x tends asymptotically to the value (Ys_0+x_0). Monod (1942) showed that the expression was in excellent agreement with data for the growth of *Escherichia coli*.

4.5 Variations in the growth curve of a simple batch culture

The growth curve shown in Fig. 4.1 is an idealized form. The duration of each growth phase may vary from practically zero to days. Information obtained from such curves is one of the foundations of cell biology.

Three common variations in the growth curve are shown in Fig. 4.2. When bacterial mass is measured by turbidity a slight decrease in the opacity is often observed on entry into the stationary phase (Fig. 4.2a). This may be an artifact reflecting change in the ratio of biomass: opacity. Another variation, depicted in Fig. 4.2(b), is a more rapid increase in the number of organisms during the first doubling. Such a result may indicate synchrony in the division of a population of unicellular organisms. The division normally becomes

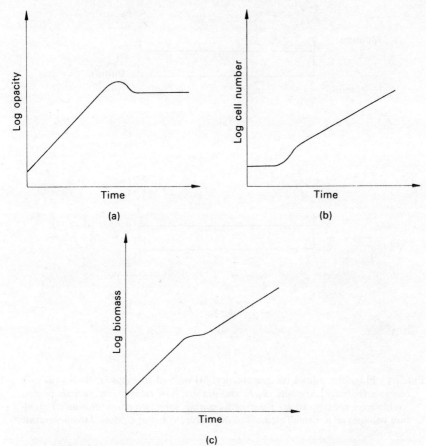

FIG. 4.2 Some observed variations in the batch growth curve: (a) fall in opacity before entry of bacterial culture into stationary phase; (b) synchronized division of
cells after lag; (c) diauxie.

completely asynchronous after about two generations. The third variation
(Fig. 4.2c), termed *diauxie*, indicates sequential utilization of substrates, the
classical case being utilization of glucose and then lactose by *Escherichia coli*.
The inflexion in the curve or even a decline in biomass occurs when the first
substrate is exhausted and a new enzyme system, which attacks the second
substrate, is induced. Study of this phenomenon eventually led to elucidation
of the mechanism of protein synthesis.

4.6 Plug-flow culture

4.6.1 General principle

A plug-flow culture (Fig. 4.3a) is a means of simulating a batch culture in an
open system. Ideally, the inoculum and the medium are mixed on entry into

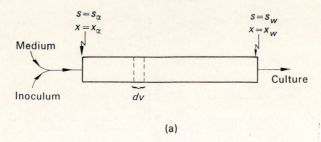

(a)

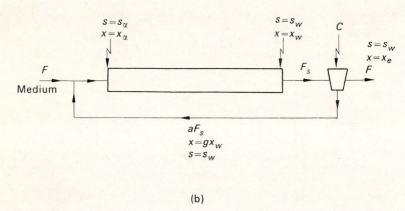

(b)

FIG. 4.3 Plug-flow culture (diagrammatic): (a) without biomass feedback; (b) with biomass feedback. *Symbols: F, F_s and aF_s are flow rates at the various points; x = biomass concentration; s = growth-limiting substrate concentration; V = culture volume; dv is a small element of the culture volume; C, device to concentrate biomass.*

the system, then the culture flows at constant velocity through the fermenter without mixing. Thus all elements of the culture should take the same 'residence' time to pass through the vessel. If $V(l)$ = volume of culture vessel, $F(l/h)$ = rate of culture flow through the fermenter then the culture residence time or 'replacement' time is given by $t_r = V/F$. In plug-flow cultures inevitably some mixing occurs and there is a velocity gradient across the vessel with fastest flow in the centre. If some mixing occurs during the passage of the culture along the fermenter then there will be a Gaussian distribution of residence times about the mean, t_r, as shown in Fig. 4.4.

4.6.2 Without biomass feedback

This system is represented in Fig. 4.3(a). When the initial concentration of growth-limiting substrate $S_\alpha \gg K_s$, growth of the biomass in each small ele-

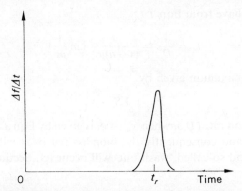

FIG. 4.4 Distribution of residence times in a near ideal plug-flow culture; Δf is the fraction of a small volume of material injected initially, which will emerge from the vessel in each small time interval, Δt.

ment of culture (dv in Fig. 4.3a) can be represented by the batch growth equation with $\mu = \mu_m$ until the growth-limiting substrate is exhausted, that is

$$x = x_\alpha \, e^{\mu_m t} \qquad 4.8$$

where x_α = initial biomass concentration and x is the concentration at time, t. If v = culture volume displaced from fermenter in time, t, then $t = v/F = v/VD$ where D is the dilution rate given by F/V. Substituting for t in Eqn 4.8 we obtain

$$\ln x = \ln x_\alpha + v\mu_m/DV \qquad 4.9$$

4.6.3 With biomass feedback

The plug-flow fermenter can be made independent of an external source of inoculum if some of the biomass is fed back to the inlet of the fermenter as depicted in Fig. 4.3b. The total flow out of the fermenter is given by

$$F_s = F + aF_s \qquad 4.10$$

where a represents the fraction of the total medium flow out of the fermenter which is fed back, hence,

$$F_s = F/(1-a) \qquad 4.11$$

Suppose the biomass is concentrated by a factor (g) before being fed back, then the fraction of the biomass which is fed back is ag. In the steady state $x_\alpha = agx_w$ where x_w is the final biomass concentration. The time for a volume v of culture to be displaced is given by $t = v(1-a)/F$. Now putting $v = V_e$ where exhaustion of growth-limiting substrate occurs, and substituting for x and t in Eqn 4.8 we have for the steady state

$$V_e = \frac{F}{(1-a)\mu_m} \left(\ln \frac{1}{ag} \right) \qquad 4.12$$

When $V_e = V$ we have from Eqn 4.12

$$t_r = \frac{1}{D_c} = \frac{1}{(1-a)\mu_m} \left(\ln \frac{1}{ag} \right)$$

4.13

and x_w is at the maximum given by

$$x_m = YS_\alpha + x_\alpha$$

4.14

The critical dilution rate (D_c), when $x_w \rightarrow 0$, is given by Eqn 4.13. If D exceeds D_c then $x_w < x_m$ and consequently the biomass fed back will be insufficient to maintain x_w and so-called 'wash out' will occur (cf. Section 6.5).

4.7 Applications of plug-flow culture

Plug-flow culture can only simulate a batch culture and no new control over the environment is introduced by the method. A novel feature of plug-flow culture is that the phases of batch culture, which are separated in time, are separated spatially in plug-flow culture.

On the laboratory scale it has so far proved impossible to realize true plug-flow culture. The main difficulties are the occurrence of laminar flow in tubular vessels so that there is a gradient of culture across the tube. Also the slow flow rate at the vessel walls encourages adhesion of the biomass to the walls of the tube. If aeration is required some mixing is inevitable. A good approximation to the plug-flow system is obtained by a number of chemostats in series (Section 6.5).

Plug-flow culture is used on the large scale in the activated sludge process. The sludge fed back is considered to be a biomass feedback, but it has been suggested (Pirt, 1972a) that it could also be a form of concentrated substrate. The anaerobic fermentations involved in the production of dairy products such as yoghurt and cheese might be run successfully as plug-flow cultures. Another possible application of the method is to simulate the conditions of intestinal flora.

CHAPTER 5

Chemostat culture

5.1 History

Prolonging a culture of microbes by the continuous addition of fresh medium and continuous harvesting of product has been discussed for more than half a century (Řičica, 1958). There are two basic types of continuous-flow culture: one is plug-flow culture and the other is the chemostat. In plug-flow culture (Section 4.6) ideally the culture travels along a tube or channel without mixing. A chemostat culture should consist of a perfectly mixed suspension of biomass into which medium is fed at a constant rate and the culture is harvested at the same rate so that the culture volume remains constant (Fig. 5.1).

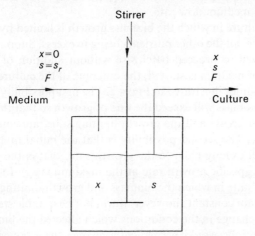

FIG. 5.1   The chemostat (diagrammatic). The biomass and growth-limiting substrate concentrations at different points are represented by x and s respectively; $F =$ flow rate; $V =$ culture volume.

The plug-flow system simulates a batch culture and offers no means of environmental control not applicable in a batch culture. The fundamental importance of chemostat culture became apparent only after the formulation of the basic theory by Monod (1950) and Novick & Szilard (1950). The theory

29

first indicated that it should be possible to fix the specific growth rate of the biomass at any value from zero to the maximum. This conclusion broke a barrier in traditional thinking, which implicitly assumed, at least in the case of bacteria, that the only stable growth rate was the maximum rate corresponding to the doubling time in a simple batch culture. Chemostat culture has revealed new horizons in microbial physiology and the history of the method shows the prime importance of the development of theory before experiment. The method is applicable to all types of protist and animal or plant tissue cells which can be grown in a homogeneous submerged culture.

5.2 Theory of the chemostat

5.2.1 General principle

The chemostat (Fig. 5.1) consists of a culture into which fresh medium is continuously introduced at a constant rate and the culture volume is kept constant by continuous removal of culture. Ideally the mixing should be perfect, that is a drop of medium entering the vessel should instantly be distributed uniformly throughout the culture. In practice this means that the time required to mix a small volume of medium with the culture should be small compared with the replacement time (t_r) given by V/F, where $V =$ culture volume and $F =$ medium flow rate.

Consider a culture in which the biomass growth is limited by the amount of a single substrate, all the other nutrients being in excess. Suppose that, at first, growth is allowed to proceed batchwise without addition of medium, then when the flow of medium is started, the outcome of the culture will be one of the three possibilities illustrated in Fig. 5.2. The first possibility is that the rate of washout of biomass will exceed the rate of growth so that the biomass concentration will decrease and the growth-limiting substrate concentration will tend towards, s_r. The second possibility is that the initial rate of washout of the biomass will exactly balance the growth rate, that is the organisms will grow with their specific growth rate at the maximum (μ_m). In this case there will be a steady state in which the biomass and growth-limiting substrate concentrations remain constant; however, it will not be a stable steady state since any temporary change in the conditions which affected the biomass and substrate concentrations would permanently change these concentrations. The third possibility would occur if the rate of washout of the biomass were initially less than the maximum growth rate. In this case the biomass concentration will continue to increase. However, eventually the decrease in the growth-limiting substrate concentration must decrease the specific growth rate until the biomass growth rate equals the wash out rate; then there can be no further changes in the concentrations of biomass and growth-limiting substrate. In this case the specific growth rate of the biomass is $<\mu_m$ and is determined by

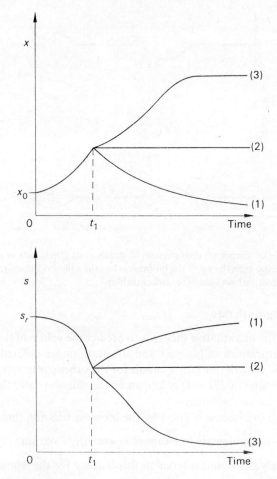

FIG. 5.2   The three possible outcomes of a chemostat culture in which growth rate of the biomass (x) is limited by the concentration of the growth-limiting nutrient (s). The flow of medium containing growth-limiting nutrient at concentration s_r is started at time, t_1. The different cases are: (1) rate of wash out of biomass exceeds maximum growth-rate; (2) rate of wash out of biomass = maximum growth rate; (3) initial rate of wash out is less than maximum growth rate of biomass.

the medium flow rate. This steady state is a self-regulating one. The consequences of temporary disturbances of the steady-state conditions are illustrated in Fig. 5.3. A fall in the biomass concentration will be associated with a rise in the substrate concentration; this will increase the growth rate and act so as to restore the steady-state conditions. A rise in biomass concentration will have the converse effect.

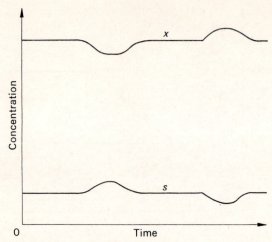

FIG. 5.3   Effects of temporary disturbances of steady-state conditions in a chemostat when the specific growth rate of the biomass is less than the maximum rate; $x =$ biomass concentration; $s =$ substrate concentration.

5.2.2 Specific growth rate

The object of the quantitative theory is to predict the values of the growth rate and the concentrations of biomass and substrate under different conditions. Let $\mu =$ specific growth rate; the symbols for the other parameters are given in Fig. 5.1. The value of $F/V = D$ is known as the dilution rate; this represents the flow rate per unit volume.

The increase in biomass is given by the biomass balance, that is

$$\text{net increase in biomass} = \text{growth} - \text{output}$$

For an infinitely small time interval dt this balance for the whole culture is

$$V.dx = V.\mu x.dt - Fx.dt \qquad\qquad 5.1$$

Dividing throughout by $V.dt$ Eqn 5.1 becomes

$$dx/dt = (\mu - D)x \qquad\qquad 5.2$$

In the steady state, when $dx/dt = 0$, we have $\mu = D$.

5.2.3 Biomass and growth-limiting substrate concentrations

The balance for the growth-limiting substrate is

$$\text{net increase} = \text{input} - \text{output} - \text{substrate used for growth}$$

For an infinitely small time interval, dt, this balance for the whole culture is

$$V.ds = F.s_r.dt - Fs - V.\mu x.dt/Y \qquad\qquad 5.3$$

where Y is the growth yield, hence

$$ds/dt = D(s_r - s) - \mu x/Y \qquad 5.4$$

In the steady state, $dx/dt = ds/dt = 0$, then the steady-state values of x and s are given by

$$(\mu - D)\tilde{x} = 0 \qquad 5.5$$

and

$$D(s_r - \tilde{s}) - \mu \tilde{x}/Y = 0 \qquad 5.6$$

where the tilde denotes a steady-state value. To obtain $\tilde{x}$ and $\tilde{s}$ we substitute for the specific growth rate

$$\mu = \mu_m s/(s + K_s) \qquad 5.7$$

Substituting $\mu = D$ in Eqn 5.7 we obtain for the steady state

$$\tilde{s} = K_s D/(\mu_m - D) \qquad 5.8$$

and from Eqn 5.6, if we put $\mu = D$, we obtain

$$\tilde{x} = Y(s_r - \tilde{s}) = Y\{s_r - K_s D/(\mu_m - D)\} \qquad 5.9$$

5.2.4 Critical dilution rate

The maximum growth rate is obtained when $\tilde{s} = s_r$. Inserting this value in Eqn 5.7 we have

$$\mu = D_c = \mu_m s_r/(s_r + K_s) \qquad 5.10$$

where D_c is the critical dilution rate, at which value the steady-state biomass concentration is zero. If $s_r \gg K_s$ it follows from Eqn 5.10 that $D_c \approx \mu_m$. Plots of $\tilde{x}$ and $\tilde{s}$ against dilution rate with typical parameter values are shown in Fig. 5.4.

5.2.5 Determination of maximum growth rate

When $s \gg K_s$ we can put $\mu = \mu_m$ in Eqn 5.2 and on integrating obtain

$$\ln x = (\mu_m - D)t + \ln x_0 \qquad 5.11$$

If we make $D > D_c$ in the chemostat, the culture biomass decreases, or washout occurs according to Eqn 5.11 and the slope of the logarithmic plot is $(\mu_m - D)$, which gives the value of μ_m. This method was used by Pirt & Callow (1960) to determine the effect of temperature on the maximum growth rate of the mould *Penicillium chrysogenum*.

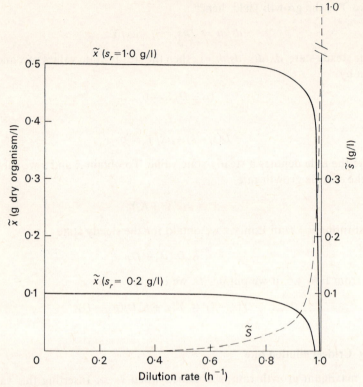

FIG. 5.4 Steady-state values of biomass ($\tilde{x}$) and growth-limiting substrate ($\tilde{s}$) concentrations in a chemostat according to Eqn 5.8 and 5.9. *Parameters:* $\mu_m = 1 \cdot 0 \text{ h}^{-1}$; $K_s = 0 \cdot 005 \text{ g/l}$; $Y = 0 \cdot 5$.

5.3 Rate of output of biomass in a chemostat

5.3.1 Determination

For a chemostat culture, the rate of output of biomass per unit volume of culture is given by $R = Dx$ and in the steady state we have

$$R = DY\{s_r - K_s D/(\mu_m - D)\} \qquad 5.12$$

The steady-state output rate as a function of D with typical parameter values is shown in Fig. 5.5. The biomass output rate reaches a maximum at dilution rate D_m, which is obtained by differentiating R with respect to D and equating the derivative to zero. Hence we find that

$$D_m = \mu_m \left\{ 1 - \left(\frac{K_s}{s_r + K_s} \right)^{1/2} \right\} \qquad 5.13$$

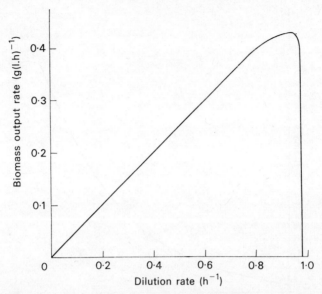

FIG. 5.5 Steady-state rates of biomass output in a chemostat. Parameters as for Fig. 5.4, $s_r = 1 \cdot 0$ g/l.

Substituting in Eqn 5.9 we obtain for the steady-state biomass at D_m,

$$\tilde{x}_m = Y[s_r + K_s - \{K_s(s_r + K_s)\}^{1/2}] \qquad 5.14$$

If $s_r \gg K_s$ then for the maximum output rate we have

$$D_m \tilde{x}_m \approx D_m Y s_r \qquad 5.15$$

5.3.2 Comparison with output of batch culture

As a consequence of autocatalysis, the maximum biomass output rate of a chemostat culture must exceed that of a batch culture of the same volume. This is shown by Fig. 5.6 where the slopes of the lines A and B represent the maximum biomass output rates. The chemostat culture can be operated continuously at the point on the growth curve where the rate of biomass production (slope B) is maximal, whereas the output rate of the batch culture is given by the mean rate of biomass production (slope A) over the whole period of the culture.

The rates of biomass output can be compared quantitatively as follows. Let x_m = maximum biomass concentration, x_0 = inoculum concentration, and let us assume that the specific growth rate in the batch culture remains at the

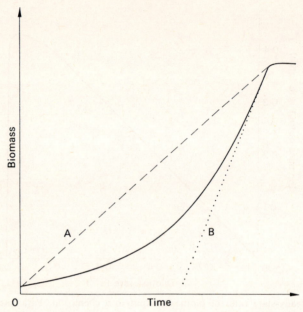

FIG. 5.6 Comparison of the maximum output rates of biomass in batch and chemostat cultures. The curve represents the growth of a batch culture and the biomass
output rate is given by the slope of the broken line, A. A chemostat culture can
maintain the output rate given by the maximum slope on the curve (dotted line, B).

maximum (μ_m) until the substrate is exhausted. Thus we have for the duration
of the batch culture

$$t_c = \frac{1}{\mu_m} \cdot \ln\left(\frac{x_m}{x_0}\right) + t_a \qquad\qquad 5.16$$

where t_a is the delay time which includes any lag period and the time required
for emptying and recharging the culture vessel between cycles. The increase in
the biomass will be Ys_r where s_r is the initial concentration of growth-limiting
substrate. Now for the output rate of the batch culture we have

$$R_{m(\text{batch})} = Ys_r/t_c = \mu_m Ys_r/\left\{\ln\left(\frac{x_m}{x_0}\right) + \mu_m t_a\right\} \qquad 5.17$$

Comparing this with the maximum output rate of a chemostat culture (given
by Eqn 5.15) we have, assuming $D_m = \mu_m$

$$\frac{R_{m(\text{chemostat})}}{R_{m(\text{batch})}} = \ln\left(\frac{x_m}{x_0}\right) + \mu_m t_a = \ln\left(\frac{x_m}{x_0}\right) + \frac{0.693 t_a}{t_d} \qquad 5.18$$

where $t_d = 0.693/\mu_m$. In practice the ratio, $x_m : x_0$ is usually 10 or greater,
which makes the maximum output rate of the chemostat at least $2.3 \times$ batch

rate, assuming that $t_a = 0$. However, frequently t_a is many times greater than t_d.

5.4 Distribution of residence times in a chemostat

Through mixing, each element of medium added may emerge at once from the culture or remain indefinitely in the culture, that is there will be a distribution of residence times. If m_0 = amount of material per unit volume present at zero time and m = amount present at time, t, then the amount of material leaving the vessel or with residence time between t and $t+dt$ will be

$$-dm = Dm.dt \qquad 5.19$$

Let the fraction of the original material with residence time between t and $t+dt$ be $-dm/m_0 = df$. Then we have from Eqn 5.19

$$df = D\frac{m}{m_0} dt \qquad 5.20$$

Since $-dm/dt = Dm$, it follows that $m/m_0 = e^{-Dt}$, hence

$$df = D e^{-Dt} dt \qquad 5.21$$

Now the fraction of material with residence times between t_1 and t_2 will be

$$F = \int_{t_1}^{t_2} D e^{-Dt} dt \qquad 5.22$$

The function, $D e^{-Dt} = df/dt$ represents the distribution of residence times.

The distribution may be determined experimentally by injecting a small volume of dye into the vessel and collecting the effluent in small fractions. The plot of the dye contents of the fractions against time gives the distribution of residence times. The exponential distribution of residence times obtained with the chemostat (Fig. 5.7) contrasts with the Gaussian distribution which is associated with plug-flow culture (Fig. 4.4). Thus the distribution of residence times characterizes each system.

Integration of Eqn 5.22 gives

$$F = e^{-Dt_1} - e^{-Dt_2} \qquad 5.23$$

Hence the fraction with residence time between 0 and t is $(1 - e^{-Dt})$. Thus it may be calculated that the fractions of the original material remaining after 1, 2, 3 and 4 replacement times respectively are 0·367, 0·135, 0·050 and 0·015. The fraction with residence time greater than t, that is from t to ∞, will be e^{-Dt}. The mean residence time of all the elements of medium, it can be shown (Denbigh, 1965, p. 82), is given by $1/D$.

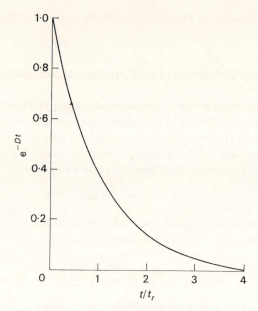

FIG. 5.7   Fraction (e^{-Dt}) of initial material remaining in a chemostat after the elapse
of different multiples of the replacement time (t_r); $t =$ time.

5.5 Deviations from chemostat theory

5.5.1 Tests of validity

The forms of the curves of biomass and growth-limiting substrate against
dilution rate constitute important tests of the validity of the theory. Excellent
agreement of experiment with the simple theory is demonstrated by the growth
of *Klebsiella aerogenes* limited by ammonium ion supply in a minimal glycerol
salts medium (Herbert, 1958). In many systems deviations from the simple
theory occur and accounting for these deviations is a basic method for determining how the biomass reacts to its environment.

It has been proposed (Contois, 1959) that the growth-limiting substrate
concentration is not solely determined by the growth rate but also increases
with biomass concentration. This deviation was accounted for by writing
$K_s = Bx$ where x is the biomass concentration and B an empirical constant.
The evidence for this deviation is doubtful in that it depended on calculation
of the limiting substrate concentrations by assuming that the growth yield
from the carbon and energy source was constant; but frequently, especially at
high substrate concentrations, the growth yield is found to decrease (see for
example Pirt, 1957); this could account for the deviation Contois observed.

Some deviations may be caused by the presence of inhibitors or stimulators of growth (Chapter 17). A deviation frequently observed is that the growth yield is not constant but depends on the growth rate. Maintenance energy (Section **8.3**) causes one such deviation; other deviations which reflect the roles of different nutrients are given in Chapter 12.

5.5.2 Effect of imperfect mixing

Good mixing in a chemostat culture implies that, on addition of a small volume of material, the time taken for the material to become homogeneously dispersed throughout the culture should be small compared with the mean residence time, that is $1/D$.

In laboratory-scale apparatus, with culture vessels of a few litres capacity, near perfect mixing can be attained with normal agitation unless the vessel contains pockets which screen the culture from the agitation, the biomass adheres to the vessel surface or the culture becomes very viscous (Section **10.2**).

If mixing is imperfect there will be spaces in the vessel in which the dilution rates are either less or greater than the mean dilution rate (D) given by F/V. This means that the dilution rates at various points in the vessel will be distributed about the mean, D. Consequently when D reaches the critical value D_c there will be some points in the vessel where $D < D_c$ and steady states can be obtained when $D > D_c$. Such a deviation from the theoretical behaviour has been termed an 'apparatus effect' (Herbert, 1958).

5.5.3 Wall growth

Many microorganisms can adhere to glass and metal surfaces (Topiwala & Hamer, 1971) and in continuous-flow cultures of long duration there may be so-called wall growth which may vary from a light film of biomass to a massive accretion of it on the vessel surface.

The effect of reproduction of the biomass in wall growth is represented in the following model (Topiwala & Hamer, 1971).

Let $\tilde{x}$ = steady-state concentration of biomass in suspension; x_w = amount of growing biomass attached to the vessel surfaces (wall growth) per unit volume of culture. In the steady state the biomass and growth-limiting substrate balances are:

$$D\tilde{x} = \mu\tilde{x} + \mu\tilde{x}_w \qquad\qquad 5.24$$

and

$$D(s_r - \tilde{s}) = (\mu\tilde{x} + \mu x_w)/Y \qquad\qquad 5.25$$

From Eqn *5.24* and *5.25* we obtain

$$\tilde{x} = Y(s_r - \tilde{s}) \qquad\qquad 5.26$$

Now substituting for $\tilde{x}$ in Eqn 5.25 and putting $\mu = \mu_m s/(s+K_s)$, we obtain

$$\frac{\mu_m \tilde{s}}{\tilde{s}+K_s} = \frac{D\,Y(s_r - \tilde{s})}{Y(s_r - \tilde{s}) + x_w} \qquad 5.27$$

Equation 5.27 is a quadratic in s, however, when $D < \mu_m$ one of the two roots is negative and only the positive root need be considered. The effect of wall growth on the steady state concentration of biomass is shown in Fig. 5.8.

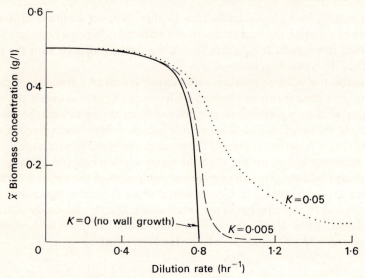

FIG. 5.8   The effect of wall growth on the steady-state biomass concentration in a chemostat culture (from Topiwala & Hamer, 1971); $Y=0.5$, $\mu_m=0.8$ h^{-1}, $s_r=$ 1·0 g/l, $K_s=0.02$ g/l, $K=x_w$ (g/l).

To maintain effectively homogeneous conditions in a chemostat culture, wall growth must be minimized. Vigorous agitation often prevents wall growth, but splashes of the culture can cause accretion of biomass above the liquid level. The adhesion of biomass can be temporarily prevented by siliconing the surface of the vessel. Teflon provides a permanent non-stick surface for biomass cultures, but this is at the moment limited to use of Teflon tubes and probes sheathed with Teflon tubes, since the problems of coating vessel surfaces with Teflon have not satisfactorily been solved.

5.6 Duration of transient states after a step change in dilution rate

The time required for a new steady state to be reached after a step change in dilution rate from D_1 to D_2 can approximately be calculated from Eqn 5.4 by putting $(s_r - \tilde{s}) \approx s_r$, $x/Y \approx s_r$ and $\mu = \mu_a$ where $\mu_a \approx (D_1 + D_2)/2$, then

$$ds/dt \approx (D_2 - D_1)s_r/2 \qquad 5.28$$

and the time required to change the steady-state concentration of growth-limiting substrate from s_1 to s_2 is given by $2(s_2-s_1)/(D_2-D_1)s_r$. This approximate formula shows that for the processes in which $\bar{s} \ll s_r$ the time for adjustment of x and s to any change in dilution rate will be only a fraction of the doubling time of the biomass. Mateles *et al.* (1965) have systematically studied the response time of an ammonia-limited culture of *Escherichia coli* to a step change in the dilution rate and found that it was insignificant for a change of $0 \cdot 1$ h^{-1}, but when D was increased by $0 \cdot 4$ h^{-1} the time required for the adjustment was about three doubling times. This indicates that the organism cannot adjust its growth rate to a considerable shift up in substrate concentration without a lag period. This lag must reflect the time required for the organism to alter its nucleic acid, enzyme content and other structures to the new steady-state levels, as suggested by theory (Chapter 24). In practice, it is advisable to make changes in dilution rate or other conditions in small steps, and allow a total time of about $3t_r$ for a new steady state to be reached.

5.7 Special purposes of chemostat culture

The three unique purposes of chemostat culture in the control of growth and behaviour of microorganisms are formulated below.

1. The chemostat permits biomass growth rate to be varied with no change in environment other than the concentration of growth-limiting substrate. In simple batch culture, growth rate changes can only be achieved by qualitative changes in nutrition or quantitative changes in physicochemical conditions such as temperature or pH value. Such methods of causing growth rate change introduce other effects which can mask the growth rate effect, for example, change in temperature can independently affect growth rate and RNA content of bacteria (Section 13.6).
2. The second purpose is the converse of (1), that is to fix the growth rate while the environment is altered. This purpose is essential to distinguish between the effects of growth rate change and environment change.
3. The third purpose is to maintain substrate-limited growth with a constant growth rate. Substrate-limited growth can be obtained only transiently in a batch culture and it is always accompanied by changing growth rate. This function of the chemostat widens the possible range of constant environments to include not only the extremes of surfeit and exhaustion of growth-limiting substrate but also all the intermediate states.

Essentially the chemostat method simplifies the culture system and thereby facilitates the elucidation of the reaction of the organism to its environment, and the control of a microbial process. The advantages of this simplification are enhanced in importance when interactions of two or more species in a culture are to be investigated or controlled.

CHAPTER 6

Elaborations on the chemostat

6.1 Introduction

Elaborations on the chemostat have increased the possible degree of control over the culture, particularly with extreme conditions as when growth rate is near the maximum or with very dilute substrates. Bryson (1952) originally devised the *turbidostat*, which is a chemostat provided with a photoelectric cell for sensing the turbidity of a culture and adding more medium when the biomass density rises above a chosen level. Turbidity measurements are limited to cultures of unicellular organisms. However, other methods of biomass sensing are now used to extend the applicability of the method, although the generic term turbidostatic control is retained for all the methods. *Biomass feedback* is a term used to refer to concentration of the biomass in the culture vessel. Thus the biomass concentration is raised above the maximum ($\approx Ys_r$) permitted without feedback. Chemostats in series form another elaboration which offers advantages in certain systems over the single chemostat. Theoretical models for these systems have been developed, but they have not as yet been tested widely probably because experimental workers have been deterred by the elaborations of culture equipment necessary to realize the systems.

6.2 Turbidostat

6.2.1 Principle

Turbidostat control with a photoelectric cell to sense biomass density is depicted diagrammatically in Fig. 6.1. The photocell signal operates the medium pump when the culture opacity exceeds the chosen value. The culture volume is kept constant by means of some constant level device. Eventually a steady state should be reached in accordance with the relation $\bar{x} = Y(s_r - \bar{s})$, that is Eqn 5.9. By means of turbidostat control, therefore, the biomass density is set and the dilution rate adjusts itself to the steady-state value, in contrast to the simple chemostat in which the dilution rate is fixed and the biomass concentration adjusts itself to the steady-state level.

42

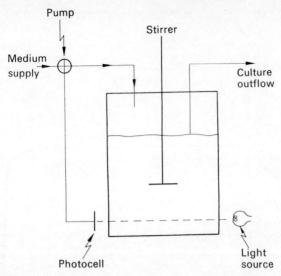

FIG. 6.1   Turbidostat (diagrammatic) with automatic control of culture opacity.

6.2.2 Means of control

Turbidity control by means of a photoelectric cell is stable when the biomass concentration varies rapidly with change in dilution rate (D), that is, near to the critical dilution rate. Where the biomass varies little with D, as shown over most of the growth rate range in Fig. 5.4, turbidity control is, in practice, too insensitive to maintain a constant dilution rate, and simple chemostat control is preferable. Technically, turbidity measurement is limited to unicellular organisms and is difficult for long-term cultures, largely because of adhesion of organisms to the surfaces of the optical cell.

Turbidostatic control may also be based on any means for sensing a growth-linked product such as carbon dioxide or acid production, or on uptake of a substrate used for growth. Watson (1972) achieved turbidostat control by means of a sensor for carbon dioxide in the gas phase. Since carbon dioxide production is linked to growth, at a fixed gas flow rate the partial pressure of carbon dioxide is given by $(P_{CO_2}) = kDx$ where k is a constant. A graph of the relation of P_{CO_2} against D for some typical data is shown in Fig. 6.2. In contrast to turbidity measurement there are two possible steady-state values for a given value of P_{CO_2}. If the controller is set to activate the medium pump only when the P_{CO_2} exceeds the set value the stable steady state is at the higher dilution rate. Conversely, if the controller is set to actuate the medium pump only when the carbon dioxide pressure falls below the set value, provided the P_{CO_2} is less than the peak value, the stable steady state is at the lower dilution rate. Thus, unlike turbidity control, P_{CO_2} control permits stable control over the

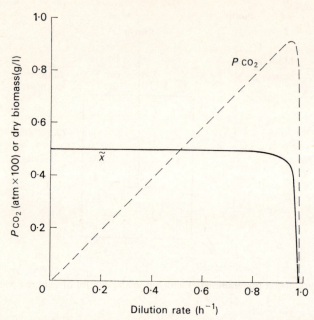

FIG. 6.2   Carbon dioxide partial pressure (P_{CO_2}) in effluent gas from chemostat as a
function of dilution rate. Data for biomass production is the same as that given in
Fig. 5.4; growth-limiting substrate concentration in feed, 1 g/l; gas flow rate 30 l
(l culture h)$^{-1}$; $\tilde{x}$ = biomass concentration.

whole range of steady states. Production of acids, such as lactic acid, could be
sensed by a pH electrode and used as a means of turbidostatic control. Also
substrate concentrations which can be determined instantly may be used as a
means of turbidostatic control; dissolved oxygen measurement by the oxygen
electrode has been applied in this way (Hospodka, 1966). Uptake of a gaseous
substrate such as oxygen or methane from the gas phase could also, in principle, be used for turbidostatic control. This method would be similar to that
based on carbon dioxide production except that the partial pressure of the
substrate gas would vary in the opposite sense to that of carbon dioxide.

6.2.3 Intracellular control of growth rate

There has been some controversy as to what process controls the biomass
growth rate when it is maintained at or close to the critical dilution rate, which
is an aim of turbidity control (Herbert, 1958). When the concentration of
growth-limiting substrate is near the K_s value, it is likely that the uptake of
growth-limiting substrate is the rate-limiting process, or 'master' reaction.
However, when the concentration of growth-limiting substrate is well above
the K_s value, it appears that the growth rate becomes virtually independent of

the substrate concentration and so no single substrate uptake acts as a master reaction. Then the growth rate may reflect the rates of many reactions in the cell. Such behaviour is indicated by a simple model of a complex reaction sequence (Dean & Hinshelwood, 1966, p. 125).

6.2.4 Applications

The turbidostat provides a means of maintaining cultures for many generations in a constant environment with excess of substrates. Since the growth rate is not fixed the system will select for faster growing organisms. Bryson (1952) originally used the method for the automatic selection of organisms resistant to antibiotics. Increase in the maximum growth rate will result from both selection of genetically different organisms and an optimization of the cellular growth mechanism, which is an expected property of an autosynthetic system (Chapter 24). Turbidostatic control should also make chemostat cultures more stable with inhibitory substrates such as phenol (Section 17.8), although a two-stage process can also be used for this purpose (Section 6.4.3).

6.3 Chemostat with feedback of biomass

6.3.1 General description

A chemostat fitted with some device to increase the biomass concentration above the value possible in the simple chemostat, that is $Y(s_r - \tilde{s})$, is termed a chemostat with feedback of biomass (Herbert, 1961; Pirt & Kurowski, 1970). The concentration may be achieved in the various ways depicted in Fig. 6.3. In system (a) a cell-free or dilute biomass stream is removed by filtration and there is also an outlet for the concentrated biomass suspension. In system (b) the culture is divided into two zones, an agitated homogeneous zone at the base where the growth occurs and, above the baffle plate, a sedimentation zone, virtually without growth, in which the biomass can sediment and return to the agitated zone. Thus a cell-free or dilute stream of biomass leaves the top of the fermenter and the concentrated biomass suspension leaves from an outlet in the growth zone. In the 'monostream' system (c) it is assumed that the biomass is concentrated by filtration or sedimentation in the outlet stream and there is no outlet for the concentrated biomass. In system (d) the biomass is concentrated outside the fermenter and the culture is separated into two streams of biomass one dilute and one concentrated; a part of the concentrated stream is fed back to the fermenter. Systems (a), (b) and (c) are represented by the *internal feedback* model and system (d) by the *external feedback* model.

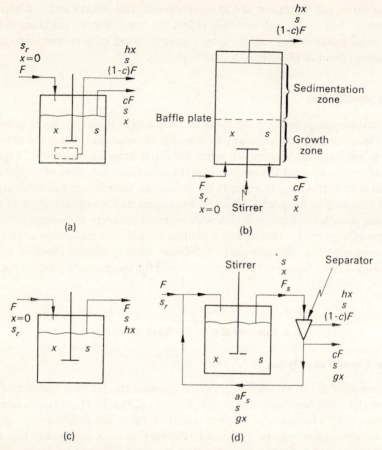

FIG. 6.3   Various systems for feedback of biomass in a chemostat: (a) internal filtration; (b) internal sedimentation; (c) monostream feedback; (d) external feedback.
*Symbols: F, F_s, flow rates; x, biomass concentration; s, s_r growth-limiting sub-
strate concentrations; c, g, h, are constants (dimensionless).*

6.3.2 Internal feedback

We consider first the filtration method depicted in Fig. 6.3(a). The biomass
and growth-limiting substrate concentrations at various points in the system
are given in Fig. 6.3(a). The fraction of the outflow, which is not filtered, is c,
therefore the outflow rate of the filtered (dilute) biomass stream is $(1-c)F$.
The concentration of biomass in the dilute stream is hx. The overall dilution
rate is given by $F/V = D$ where V is the culture volume. The biomass balance
for the culture is

net rate of   = growth rate − rate of output in − rate of output in
increase   concentrated stream   dilute stream

which, for unit volume of culture is

$$dx/dt = \mu x - cDx - (1-c)Dhx \qquad 6.1$$

that is

$$dx/dt = [\mu - D\{c(1-h)+h\}]x \qquad 6.2$$

In the steady state when $dx/dt = 0$ we have

$$\mu = AD \qquad 6.3$$

where $A = c(1-h)+h$. When the filtered effluent is cell-free, that is $h=0$, then $A=c$. On the other hand when the filtration removes very little biomass, that is $h \rightarrow 1$, then we may write $A=h$. Thus it follows that the limits of A are c to h. If $h=1$ then there is no feedback. With feedback, it follows that $\mu < D$.

Since the specific growth rate is given by $\mu = \mu_m s/(s+K_s)$, and, for the steady state, on substituting for μ, we have

$$AD = \mu_m \tilde{s}/(\tilde{s}+K_s) \qquad 6.4$$

hence

$$\tilde{s} = AK_s D/(\mu_m - AD) \qquad 6.5$$

The growth-limiting substrate balance becomes the same as that for the simple chemostat (Eqn 5.4) hence, in the steady state, when $\mu = AD$ we have

$$\tilde{x} = (s_r - \tilde{s})Y/A \qquad 6.6$$

From Eqn 6.6 we see that feedback increases the biomass concentration by the factor $1/A$, which is called the 'concentration' factor. At the critical dilution rate (D_c), $s=s_r$ and, when $s_r \gg K_s$, Eqn 6.4 shows that with respect to the simple chemostat the critical dilution rate is increased to the value μ_m/A.

The biomass output rate per unit volume of culture in steady states is given by

$$R = \{(1-c)h+c\}D\tilde{x} = AD\tilde{x} \qquad 6.7$$

The effects of biomass feedback on the biomass concentration and output rate are depicted in Fig. 6.4. An important effect is that feedback increases the maximum biomass output rate of a culture. Biomass feedback by filtration in a chemostat was experimentally realized by Pirt & Kurowski (1970).

Biomass feedback may also be achieved by means of sedimentation as shown in Fig. 6.3(b). If it is assumed that growth occurs only in the homogeneous zone below the baffle plate and that the sedimented biomass is immediately capable of growth on return to the growth zone, then the growth of such a culture should accord with the above model for internal feedback.

In the monostream system of internal feedback (Fig. 6.3c) there is an outlet for a dilute biomass stream only. This requires some sedimentation or filtration of the biomass in the outlet stream. For this system we put $V =$ volume of

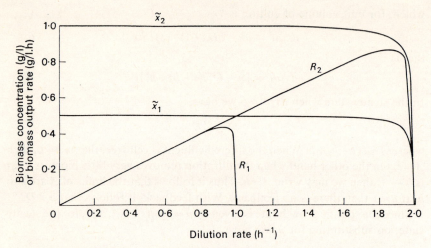

FIG. 6.4 Comparison of biomass concentrations and output rates in steady states of chemostat cultures with and without feedback. *Symbols:* $\tilde{x}_1$ = biomass concentration in chemostat without feedback; $\tilde{x}_2$ = biomass concentration in chemostat culture with feedback; R_1 = biomass output rate per unit volume without feedback, R_2 = biomass output rate of chemostat with feedback; $\mu_m = 1\cdot00$ h^{-1}; $s_r = 1\cdot0$ g/l; $K_s = 0\cdot005$ g/l; 'concentration factor' (A or B) = 2$\cdot$0.

culture in the growth zone. This system is represented by the model for internal feedback in which $c = 0$ and $A = h$, if a steady state can be realized. The tower fermenter as described by Royston (1966) is of this type but, in practice, it seems impossible to control the concentration factor and achieve steady states. Hence it is necessary to use the dual stream system with outlets for dilute and concentrated biomass to be able to control the concentration factor. Internal feedback may inadvertently be realized in a simple chemostat if the outlet tube permits some of the biomass to sediment in the fermenter, or if some obstruction in the outlet tube acts as a biomass filter.

6.3.3 External feedback

The chemostat with external feedback of biomass and its parameters are shown in Fig. 6.3(d). The effluent culture is passed through a separator, for example a centrifuge, which concentrates the biomass and produces both dilute and concentrated streams of biomass. Part of the concentrated stream is fed back to the culture vessel. The overall dilution rate is given by $F/V = D$ where V is the culture volume. The culture outflow rate from the fermenter is given by

$$F_s = F + aF_s \qquad\qquad 6.8$$

where a is the fraction of the outflow liquid stream which is fed back. From

Eqn 6.8 it follows that $F_s = aF/(1-a)$. The separator concentrates the biomass by the factor g and since the amount of biomass fed back is agF_s, it can be seen that ag is the fraction of the biomass leaving the fermenter, which is fed back.

The biomass balance for the culture is given by

$$\text{net growth} = \text{growth} - \text{output} + \text{feedback}$$

which for the whole culture will be

$$V.dx = V\mu x.dt - F_s x.dt + aF_s gx.dt \qquad 6.9$$

Substituting for F_s and dividing through by $V.dt$, Eqn 6.9 becomes

$$dx/dt = \mu x - Dx/(1-a) + agDx/(1-a) \qquad 6.10$$

and in the steady state when $dx/dt = 0$ we have

$$(\mu - BD)\tilde{x} = 0 \qquad 6.11$$

where $B = (1-ag)/(1-a)$. The factor B is a positive fraction because $1 > ag > a$. If $ag = 1$ it would mean that all the biomass is fed back and there could not be a steady state. In the steady state $\mu = BD$, thus $\mu < D$.

Substituting $\mu = \mu_m s/(s+K_s)$ we obtain

$$\tilde{s} = BDK_s/(\mu_m - BD) \qquad 6.12$$

For the growth-limiting substrate balance we have

$$\begin{matrix} \text{net rate of} = & \text{input} + & \text{feedback} - & \text{output} - & \text{substrate utilized} \\ \text{increase} & \text{rate} & \text{rate} & \text{rate} & \text{for growth} \end{matrix}$$

that is, for unit volume

$$ds/dt = Ds_r + aDs/(1-a) - Ds/(1-a) - \mu x/Y \qquad 6.13$$

which reduces to the equation for ds/dt in the simple chemostat (Eqn 5.4). Substituting $\mu = BD$ in Eqn 5.4 we obtain for the steady state value of x

$$\tilde{x} = (s_r - \tilde{s})Y/B \qquad 6.14$$

where $1/B$ is the concentration factor. The critical dilution rate (D_c) will occur when $\tilde{s} = s_r$ and if $s_r \gg K_s$, $D_c \approx \mu_m/B$.

The concentration of the biomass (hx) in the dilute biomass effluent is obtained from the biomass balance

$$F_s\tilde{x} = aF_s g\tilde{x} + cFg\tilde{x} + (1-c)Fh\tilde{x} \qquad 6.15$$

and, substituting for F_s we obtain

$$h = (B-cg)/(1-c) \qquad 6.16$$

The rate of output of the biomass per unit volume of the system will be

$$R = F_s\tilde{x}/V - aF_s g\tilde{x}/V \qquad 6.17$$

and on substituting for F_s we find

$$R = BD\tilde{x} \qquad\qquad 6.18$$

6.3.4 Advantages of biomass feedback

By means of biomass feedback, the maximum output rate of biomass and products in a chemostat with a given medium can be increased. The increase will be α fold where α is the concentration factor. This is useful when the growth-limiting substrate is unavoidably dilute, for example, in brewing, or effluent purification, or where the substrate has a low solubility. In beer brewing and sewage purification a concentration factor of 100 is feasible. The system is advantageous also when the concentration of growth-limiting substrate has to be limited because of the formation of an inhibitory product. Such a system appears in the reduction of ferrous to ferric ions by *Ferrobacillus* species (D. Kelly, private communication). Also addition of biomass feedback will protect against 'shock loading' with an inhibitory substrate because the critical dilution rate is raised.

6.4 Chemostats in series

6.4.1 Multi-stream type

The joining together of two or more chemostats in series (Fig. 6.5) leads to a multi-stage process which may have different conditions in each of the stages. We will consider a two-stage system of the type depicted in Fig. 6.5. This is

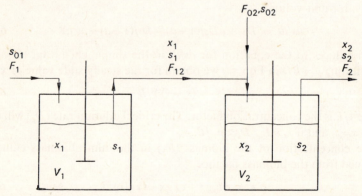

FIG. 6.5 Two chemostats in a series of 'multi-stream' type. The symbols F, V, x and s represent respectively the flow rates, culture volumes, biomass and growth-limiting substrate concentrations at the various points. In the 'single-stream' type the flow of medium (F_{02}) into the second stage would be zero.

called the multi-stream type since it has medium feeds to both first and second stages. The following analysis of the system is based on that of Herbert (1964).

Let F_{01} = rate of flow of medium to stage 1; F_{12} = rate of flow of culture from stage 1 to stage 2; F_{02} = rate of flow of medium to stage 2; the culture volumes in first and second stages are V_1 and V_2 respectively. The overall dilution rate in the second stage is given by

$$D_2 = (F_{02}+F_{12})/V_2 = F_{02}/V_2+F_{12}/V_2 = D_{02}+D_{12} \qquad 6.19$$

The terms D_{02} and D_{12} are termed partial dilution rates. The biomass balance in the second stage is given by

net rate of = growth rate + input rate − output rate
increase

This balance for unit volume during the small time interval dt is

$$dx_2/dt = \mu_2 x_2 + D_{12}x_1 - D_2 x_2 \qquad 6.20$$

where x_1 and x_2 are the biomass concentrations in first and second stages respectively. In the steady state when $dx_2/dt = 0$, Eqn 6.20 becomes

$$(\mu_2 - D_2)\tilde{x}_2 + D_{12}\tilde{x}_1 = 0 \qquad 6.21$$

hence

$$\mu_2 = D_2 - D_{12}\tilde{x}_1/\tilde{x}_2 \qquad 6.22$$

From Eqn 6.22 it follows that $\mu_2 < D_2$. Rearrangement of Eqn 6.22 gives

$$\tilde{x}_2 = D_{12}\tilde{x}_1/(D_2 - \mu_2) \qquad 6.23$$

hence as long as $\tilde{x}_1$ is finite, $\tilde{x}_2$ is finite, no matter how great D_2 is; this means that there is no critical dilution rate for the second stage.

The growth-limiting substrate balance for the second stage is

rate of = rate of input + rate of input − outflow − consumption
increase   from first   of fresh   rate   rate
             stage   medium

that is, for unit volume,

$$ds_2/dt = D_{12}s_1 + D_{02}s_{02} - D_2 s_2 - \mu_2 \tilde{x}_2/Y \qquad 6.24$$

In the steady state Eqn 6.24 becomes

$$D_{12}\tilde{s}_1 + D_{02}s_{02} - D_2 \tilde{s}_2 - \mu_2 \tilde{x}_2/Y = 0 \qquad 6.25$$

If we substitute for μ_2 in Eqn 6.25 by means of Eqn 6.22 and put $\tilde{x}_1 = Y(s_{01}-\tilde{s}_1)$ then we obtain

$$\tilde{x}_2 = Y\left(\frac{D_{12}}{D_2}s_{01}+\frac{D_{02}}{D_2}s_{02}-\tilde{s}_2\right) \qquad 6.26$$

To solve for $\tilde{s}_2$ we substitute in Eqn 6.25 $\mu_2 = \mu_m s_2 / (s_2 + K_s)$ then, from Eqn 6.25 and 6.26, we obtain

$$(\mu_m - D_2)\tilde{s}_2^2 - \left\{ \frac{\mu_m D_{12} s_{01}}{D_2} + \frac{(\mu_m - D_2) D_{02} s_{02}}{D_2} - D_{12}\tilde{s}_1 + K_s D_2 \right\} \tilde{s}_2$$
$$+ K_s(D_{12}\tilde{s}_1 + D_{02} s_{02}) = 0 \qquad 6.27$$

The solution is

$$\tilde{s}_2 = \{-b - (b^2 - 4ac)^{1/2}\}/2a \qquad 6.28$$

where $a = (\mu_m - D_2)$; $-b = \mu_m D_{12} s_{01}/D_2 + (\mu_m - D_2) D_{02} s_{02}/D_2 - D_{12}\tilde{s}_1 + K_s D_2$; $c = K_s(D_{12}\tilde{s}_1 + D_{02} s_{02})$. The other root of the equation is negative and has no practical significance.

The data of Fig. 6.6. show the effect of varying the rate of flow of culture

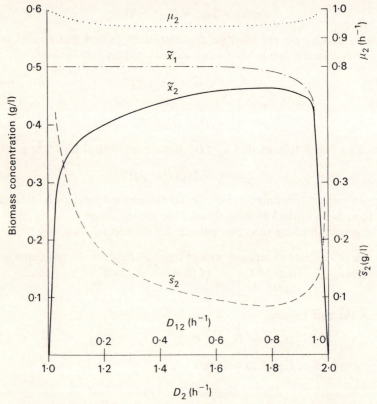

FIG. 6.6 Steady-state values of the specific growth rate (μ_2) and concentrations of biomass ($\tilde{x}_2$) and growth-limiting substrate ($\tilde{s}_2$) in the second stage of a series of two chemostats when the dilution rate in the first stage (D_1) is varied: $\tilde{x}_1 =$ concentration of biomass in first stage; $\mu_m = 1 \cdot 0$ h^{-1}; $K_s = 0 \cdot 005$ g/l; $s_{01} = s_{02} = 1 \cdot 0$ g/l; $Y = 0 \cdot 5$; $D_{02} = 1 \cdot 0$ h^{-1}; D_{12} is varied from o to $1 \cdot 0$ h^{-1}; $D_2 =$ overall dilution rate in second stage.

from the first to the second stage (D_{12}) when the rate of addition of fresh medium to the second stage (D_{02}) is fixed. The specific growth rate in the second stage is maintained nearly at the maximum ($\not< 0.94\,\mu_m$) at all dilution rates. Another feature is that over most of the dilution rate range the biomass concentration is fairly constant, that is stable, despite the high growth rate.

When the rate of flow of culture from first to second stage (D_{12}) is fixed and the feed rate of fresh medium (D_{02}) is varied the effects are as shown in Fig. 6.7. In the second stage the biomass concentration falls gradually when

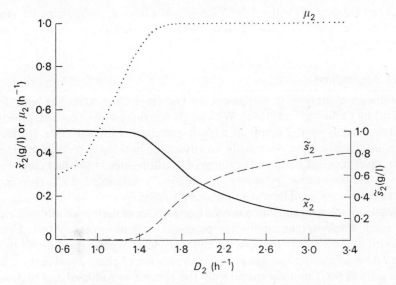

FIG. 6.7 Steady-state values of specific growth rate (μ_2) and concentrations of biomass ($\tilde{x}_2$) and growth-limiting substrate ($\tilde{s}_2$) in second stage of a series of two chemostats when the rate of flow from the first stage (D_{12}) is fixed and the rate of addition of fresh medium (D_{02}) is varied: $D_{12} = 0.5\,h^{-1}$; values of other parameters are given in Fig. 6.6.

the growth rate is virtually at the maximum, in contrast to the simple chemostat (Fig. 5.4).

It is assumed in the above treatment that there is no lag in the organism's response to the change in growth-limiting substrate concentration on passage from the first to the second stage. However, it is known that when a large shift up in growth rate is required an appreciable lag may occur before the new growth rate is attained (Section 5.6). On the other hand, a shift down in growth rate seems unlikely to result in a significant lag in adjustment of the growth rate, although the transient changes in biomass contents of RNA and other constituents may be prolonged.

6.4.2 Single stream type without feedback

In the single stream system medium is fed into the first stage only of the series of chemostats. Thus, in the two-stage system shown in Fig. 6.5, $D_{02}=0$ and $D_{12}=D_2$ so that Eqn 6.26 becomes

$$\tilde{x}_2 = Y(s_{01}-\tilde{s}_2) \qquad\qquad 6.29$$

and for the specific growth rate from Eqn 6.22 we have

$$\mu_2 = D_2(\tilde{x}_2-\tilde{x}_1)/\tilde{x}_1 \qquad\qquad 6.30$$

With most systems $\tilde{x}_2$ will not differ significantly from $\tilde{x}_1$ and $\mu_2 \approx 0$ unless the dilution rate in the first stage is near to the critical value, or there is an inhibitor present.

6.4.3 Applications

The design of laboratory equipment for two-stage chemostats has been described by Callow & Pirt (1961). With single-stream systems, in which growth is limited solely by the supply of a single growth-limiting substrate, for biomass production there is virtually no advantage in using a series of chemostats. Over most of the possible range of dilution rates in the first stage the growth-limiting substrate would be practically exhausted and stationary phase conditions would occur in the second stage.

When complex media with more than one source of carbon or nitrogen etc. are used, single-stream multi-stage processes may be necessary to achieve utilization of all substrates. For example, Harte & Webb (1967) found that when *Klebsiella aerogenes* was grown on a mixture of glucose and maltose at high growth rates in a chemostat only the glucose was utilized and maltose utilization was repressed. In the second stage of a series of two fermenters maltose utilization was completed. In general, the multi-stage system provides a series of different environments.

The multi-stream process is a valuable means for obtaining steady-state growth when in the simple chemostat the steady state is unstable, as may happen when the growth-limiting substrate is also a growth inhibitor (Section 17.8). The work of Jones et al. (1973) is a good example of the application of the two-stage process to determine bacterial growth rates with phenol as the inhibitory growth-limiting substrate. In production of secondary metabolites by chemostat culture the second stage may be used to provide a non-growing stage in which secondary metabolite production occurs (Section 16.5.4).

The two-stage system extends the range of application of chemostat culture in that the second stage may be used to extend the growth rate downwards to zero, and to achieve stable conditions with maximum growth rate, both of which conditions are impossible in the simple chemostat.

6.5 Chemostats in series with biomass feedback

A series of chemostats with feedback of biomass is depicted in Fig. 6.8. The system is provided with a biomass separator to return concentrated biomass to the first stage. This system is of interest in that, with a sufficient number of stages, it simulates a plug-flow fermenter. Powell & Lowe (1964) have

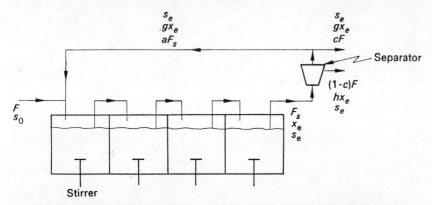

FIG. 6.8 Series of chemostats with feedback of biomass. The symbols x and s represent the concentrations of biomass and growth-limiting substrate respectively at various points; F, F_s and aF_s are the flow rates at various points; a is the fraction of the liquid flow (F_s), which is fed back $\{F_s = F/(1-a)\}$.

analysed the behaviour of the system, however, the general equations for the steady-state concentrations of biomass and growth-limiting substrate in the rth stage of the series of chemostats could not be solved analytically. The overall dilution rate is $D = F/wv$ where v is the volume of culture in each stage and w is the number of stages. The critical overall dilution rate was found by Powell & Lowe (1964) to be

$$D_c = \left\{ \frac{1-a}{(1-b^{1/w})w} \right\} \mu_m \frac{s_0}{s_0 + K_s} \qquad 6.31$$

where a is the fraction of the total liquid flow through the series of fermenters which is fed back, and b is the fraction of the biomass in the wth stage which is fed back ($b = ag$ where g is the biomass concentration factor achieved in the separator); s_0 = concentration of growth-limiting substrate in the initial medium. The limit of D_c as w is increased to infinity is obtained by inserting into Eqn 6.31 the limit

$$\lim_{w \to \infty} (1-b^{1/w})w = \log 1/b \qquad 6.32$$

hence

$$\lim_{w \to \infty} D_c = \frac{(1-a)}{\log 1/b} \frac{\mu_m s_0}{(s_0 + K_s)} \qquad 6.33$$

Equation *6.33* is the same as that for the critical dilution rate in a plug-flow fermenter (Section **4.6.3**). Powell & Lowe (1964), by a computer method, calculated the biomass concentration (x_e) obtained in the final stage of a series of five chemostats with feedback. Figure 6.9 shows the results of Powell & Lowe

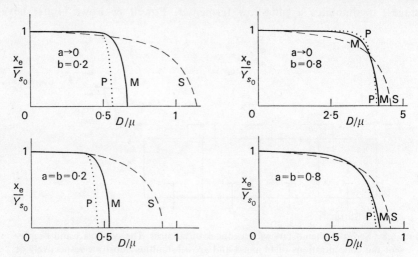

FIG. 6.9   Comparison of the steady-state values of the biomass concentrations (x_e) in the effluent from: (M) a series of five chemostats with feedback; (P) an ideal plug-flow culture with feedback; (S) a single chemostat with feedback. $\mu = \mu_m s_0/(s_0 + K_s)$ where s_0 is the concentration of growth-limiting substrate ($= 10 K_s$) in the medium added; D = overall dilution rate; Y = growth yield (from Powell & Lowe, 1964); $a \to 0$ means that a is made very small.

comparing the biomass concentrations in effluents from the series of chemostats, the ideal plug-flow culture and the single chemostat, all with biomass feedback. It is apparent that with a low a and high b more complete utilization of growth-limiting substrate can be obtained over most of the dilution rate range in the series of chemostats than is possible with the single chemostat. This advantage is greater the lower the initial substrate concentration. Also Powell & Lowe found that with low initial substrate concentration the maximum rate of output of biomass from the series of chemostats could exceed that from a single-stage chemostat with feedback. The results also show that even with only five fermenters in the series the system closely approaches the ideal plug-flow culture.

A multi-stage chemostat with feedback has been described by Kitai *et al.* (1969). The system is potentially of interest for the conversion of very dilute substrates into biomass and as a means for realizing plug-flow culture.

CHAPTER 7

Death of cells in growing cultures

7.1 Definitions of dead and dormant cells

Protists, which become incapable of growth in an environment which is normally suitable for growth, must be either dead or dormant cells. Essentially the difference between the two is that dormant cells can, under some conditions, regenerate the normal growing cell or 'vegetative' form whereas dead cells cannot. Examples of dormant forms are the spores of bacteria and fungi and the cysts of protozoa. Generally the production of dormant forms is an ordered process with a duration longer than the minimum doubling time of the vegetative form. This change of cell type is somewhat analogous to the differentiation of tissues in higher animals and plants although it may be unjustified to carry the analogy very far, and tissue cell differentiation is not considered here. A model for the process of bacterial differentiation into spores is considered in Section 18.5.

Protists may die as a result of an adverse physical factor such as high temperature, or through the effect of a toxic chemical, starvation or 'mistakes' in autosynthesis. In a growing culture dormant and dead cells behave similarly in that they make no contribution to the growth of the population.

7.2 Death rate

7.2.1 Death of cells in a batch culture

We suppose that the cell death rate is proportional to the number of viable cells (y_v) present, that is

$$dy_v/dt = -ky_v \qquad 7.1$$

where k, the specific death rate, is a constant. The rate of growth of the viable population is given by

$$dy_v/dt = (\mu - k)y_v \qquad 7.2$$

where μ is the specific growth rate, hence

$$\ln y_v = \ln y_{v(0)} + (\mu - k)t \qquad 7.3$$

The apparent specific growth rate is $(\mu - k)$ and as long as $k < \mu$ the viable population grows exponentially. The apparent doubling time is given by $(\ln 2)/(\mu - k)$ but the true doubling time is $(\ln 2)/\mu$.

The rate of increase in the total cell population (y_T) is given by

$$dy_T/dt = \mu y_v \qquad 7.4$$

Substituting for y_v from Eqn 7.3 we have

$$\int_{y_{T(0)}}^{y_T} dy_T = \mu y_{v(0)} \int_0^t e^{(\mu - k)t} \, dt \qquad 7.5$$

that is

$$y_T = y_{T(0)} + \frac{\mu}{\mu - k} \cdot y_{v(0)} (e^{(\mu - k)t} - 1) \qquad 7.6$$

For the increase in the total population we have three possible cases depending on whether $k < \mu$ or $k = \mu$ or $k > \mu$. When $k < \mu$ and t is large we can put $y_{T(0)} \approx 0$ and $(e^{(\mu - k)t} - 1) \approx e^{(\mu - k)t}$ then

$$y_T \approx \frac{\mu}{\mu - k} y_{v(0)} \, e^{(\mu - k)t} \qquad 7.7$$

that is y_T will increase at a constant exponential rate.

If $k = \mu$ then y_v is constant, and Eqn 7.6 for y_T is no longer applicable. Instead we have from Eqn 7.4

$$y_T = y_{T(0)} + \mu y_{v(0)} t \qquad 7.8$$

If $k > \mu$ then $y_v \to 0$ and from Eqn 7.6 it follows that the total population approaches a finite limit given by

$$\lim_{(t \to \infty)} y_T = y_{T(0)} + \frac{\mu}{k - \mu} \cdot y_{v(0)} \qquad 7.9$$

7.2.2 Death of cells in a chemostat culture

In a chemostat in which some cell death occurs the viable cell population balance in the small interval dt is given by

$$dy_v = \mu y_v \, dt - k y_v \, dt - D y_v \, dt \qquad 7.10$$

where k is the specific death rate and D is the dilution rate. Hence we obtain

$$dy_v/dt = \{(\mu - k) - D\} y_v \qquad 7.11$$

and, therefore, in the steady state when $dy_v/dt = 0$

$$\mu = D + k \qquad 7.12$$

From the balance for the total cell population (y_T), if $y_v = \beta y_T$ where β is the fraction viable then we have

$$\text{net growth} = \text{growth} - \text{output}$$

that is

$$dy_T = \mu\beta y_T \, dt - Dy_T \, dt \qquad 7.13$$

hence

$$dy_T/dt = (\mu\beta - D)y_T \qquad 7.14$$

and in the steady state

$$\mu = D/\beta \qquad 7.15$$

Comparing Eqn 7.15 and 7.12 we find that

$$k = D\left(\frac{1}{\beta} - 1\right) \qquad 7.16$$

Also since μ has the maximum value, μ_m, from Eqn 7.15 the fraction viable (β) has a minimum value in the steady state given by

$$\beta_{min} = D/\mu_m \qquad 7.17$$

7.3 Production of dead cells at the time of division

7.3.1 In a chemostat culture

It is sometimes found that even under optimal conditions for growth some dead cells are produced in a culture. This may be interpreted to mean that there is a certain probability of an individual cell being dead at birth, perhaps through a mistake in autosynthesis. Let us suppose that when a number (y_v) of cells grow and divide they produce $2y_v$ new cells of which a fraction θ is viable. Then $2\theta y_v$ viable cells and $2(1-\theta)y_v$ dead cells are produced with each new generation of cells. The fraction θ, called the *viability index* (Powell, 1956) represents the probability that a newly formed cell will be viable. Thus if 50 cells divide and become 100 cells of which 99 are viable, the viability index is 0·99, which is the probability that a new cell will be viable. It follows that if $\theta < 0·5$ the number of viable cells (y_v) will tend to zero; if $\theta = 0·5$ the number viable remains constant, and if $\theta > 0·5$ then the number viable will increase.

To obtain the viable cell balance we note that in a population of y_v viable organisms the number of cell divisions in the small interval dt is $\mu y_v \, dt$. The number of new viable organisms produced is $2\mu\theta y_v \, dt$ where θ is the viability index, and $\mu y_v \, dt$ old organisms are lost by the division process. The viable cell balance in a chemostat is given by

net increase = number produced − number lost through division − output

that is

$$dy_v = 2\mu y_v \theta \, dt - \mu y_v \, dt - Dy_v \, dt \qquad 7.18$$

If the fraction of the total population viable is β we can substitute $y_v = \beta y_t$ and $dy_v = \beta\, dy_t$, then we obtain

$$dy_t/dt = 2\mu\theta y_t - \mu y_t - Dy_t \qquad 7.19$$

that is

$$dy_t/dt = \{(2\theta - 1)\mu - D\} y_t \qquad 7.20$$

In the steady state when $dy_t/dt = 0$ we have

$$\mu = D/(2\theta - 1) \qquad 7.21$$

Comparing Eqn 7.21 and 7.15 it is seen that in the steady state,

$$\beta = 2\theta - 1 \qquad 7.22$$

Also as in the case of cell death by killing the minimum value of β is given by Eqn 7.17.

7.3.2 In a batch culture

The production of dead cells at division in a batch culture is modelled as follows. If θ is the viability index the number of viable cells present in a population after n generations will be

$$y_v = y_{v(0)}(2\theta)^n \qquad 7.23$$

where $y_{v(0)}$ is the initial number viable. The number of dead cells accumulated after n generations will be

$$y_d = 0 + 2y_{v(0)}(1 - \theta) + 2y_{v(0)}2\theta(1 - \theta) + 2y_{v(0)}(2\theta)^2(1 - \theta) + \cdots$$
$$+ 2y_0(2\theta)^{n-1}(1 - \theta) \qquad 7.24$$

The successive terms in Eqn 7.24 represent the increase in the dead cells at each generation, the initial number being zero. Eqn 7.24 forms a geometrical progression with constant ratio 2θ. Summing the series gives

$$y_d = 2(1 - \theta)\frac{\{1 - (2\theta)^n\}}{1 - 2\theta} \cdot y_{v(0)} \qquad 7.25$$

As $n \to \infty$ when $\theta > 0.5$, $1 - (2\theta)^n \to -(2\theta)^n$ and

$$y_d \to \frac{2(\theta - 1)(2\theta)^n}{1 - 2\theta} \cdot y_0 \qquad 7.26$$

The fraction of the total population viable after n generations will be given by

$$\beta = \frac{y_0(2\theta)^n}{y_0(2\theta)^n + y_d} \qquad 7.27$$

Substituting for y_d from Eqn 7.26 it follows that as $n \to \infty$, $\beta \to (2\theta - 1)$, that is the steady-state value obtained in a chemostat culture (Eqn 7.22).

The growth rate of the viable population is given by

$$dy_v = \mu y_v 2\theta . dt - \mu y_v . dt \qquad 7.28$$

where the first term on the right hand side gives the increase in the number of viable cells after $\mu\, dt$ cell divisions and the second term is the loss of old cells through the divisions. Hence the growth rate of the viable cell population will be

$$dy_v/dt = (2\theta - 1)\mu y_v \qquad 7.29$$

Integration of Eqn 7.29 gives

$$\ln y_v = \ln y_{v(0)} + (2\theta - 1)\mu t \qquad 7.30$$

Equation 7.30 shows that the viable population number will increase exponentially provided $\theta > 0.5$ and the apparent specific growth rate is $(2\theta - 1)\mu$, which approaches $\beta\mu$ as the number of generations increases.

The growth rate of the total cell population is given by

$$dy_t/dt = \mu\beta y_t \qquad 7.31$$

After some generations when $\theta > 0.5$, the value of β becomes nearly constant and the specific growth rate of the total population approaches the apparent specific growth rate of the viable population. The effect of production of dead cells is depicted in Fig. 7.1. It is apparent that with $\theta = 0.9$, even after only

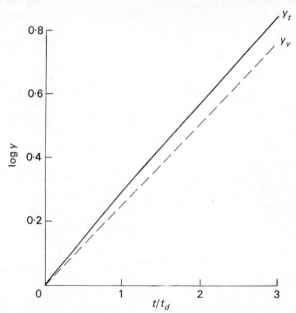

FIG. 7.1 Plot of logarithm of viable count (y_v) and total count (y_t) in a batch culture when the viability index (θ) = 0.9. The fraction viable (β) tends to 0.8. The abscissa gives t/t_d where t = time and t_d = true doubling time, that is $(\ln 2)/\mu$.

three generations the slopes of viable and dead cell growth curves are nearly the same and β has nearly reached its final value of 0·8.

7.4 Effects of cell death on growth

The foregoing analysis may be used to predict the effects of cell death on the growth of cultures. The chemostat could be a useful tool to study the effect of growth rate on death rate, however, it should be noted that the maximum value of the specific death rate in steady states is limited to $(\mu_m - D)$.

Cell death may result in autolysis of the cell or release of cell contents into the medium, which can alter the nutritional conditions and affect the metabolism and growth rate of the surviving cells (Postgate & Hunter, 1962).

CHAPTER 8

Energy and carbon source requirements

8.1 Introduction

The fungi, protozoa and most of the bacterial groups are *heterotrophic*, that is they obtain their cell carbon from organic compounds. The remainder of the protists, consisting of the algae and some of the bacterial groups, obtain their cell carbon from carbon dioxide and are called *autotrophic*. Organisms require a source of energy to synthesize cell material for growth and for so-called 'maintenance' functions. A classification of carbon and energy requirements of organisms, based on that of Stanier *et al.* (1963) p. 292), is shown in Table 8.1. The light energy requirements of the photosynthetic protists are not

TABLE 8.1 Classification of energy sources and electron donors and acceptors for protists

Type of organism	Energy source	Electron acceptors in energy-yielding process	Electron donors in energy-yielding process or for CO_2 reduction
Photolithotrophic	light	—	water, reduced sulphur compounds
Photoorganotrophic	light	—	organic compounds
Chemolithotrophic	oxidation–reduction	oxygen, carbon dioxide	inorganic (hydrogen, reduced sulphur, reduced nitrogen, ferrous ions)
Chemoorganotrophic	oxidation–reduction	oxygen, nitrate, sulphate, organic compounds	organic compounds

considered here, but have been treated by Pipes (1962) and Göbel & Pfennig (1969). Oxygen is one of the commonest oxidants in energy metabolism; in the absence of oxygen, nitrate or sulphate ions are used as electron acceptors by some bacteria. The chemolithotrophs oxidize inorganic reduced forms of nitrogen and sulphur, ferrous ions or hydrogen to obtain energy under aerobic conditions, whereas methanogenic bacteria are anaerobic autotrophs, which use hydrogen to reduce carbon dioxide to methane to obtain energy.

Other protists, particularly among the bacteria and yeasts, obtain their energy under anaerobic conditions by oxidation–reduction reactions performed on organic compounds, a process, which has been called *fermentation*.

To define quantitatively the requirements for carbon and energy sources we need to know the K_s values and growth yields from the substrates. Table 2.1 shows that the K_s values for carbon and energy sources are about 10^{-5} M, that is, the highest for any nutrient; only the K_s values for phosphate, potassium or magnesium ions are of the same order.

Representative values for the overall growth yields from carbon and energy sources are given in Table 8.2. These overall yields include the requirements

TABLE 8.2   Overall growth yields from various carbon and energy sources

Organism	Substrate	Maximum observed growth yield (g dry weight/g)		
		Substrate	Substrate carbon	Oxygen used
Candida utilis[1]	glucose	0·51	1·28	1·30
Penicillium chrysogenum[2]	glucose	0·43	1·08	1·35
Aerobacter cloacae[3]	glucose	0·44	1·10	1·03
Candida utilis[1]	acetic acid	0·36	0·90	0·62
Candida utilis[1]	ethanol	0·68	1·30	0·58
Pseudomonas sp.[5]	methanol	0·41	1·09	0·44
Bacterial sp.[6]	*n*-pentane	0·84	1·01	0·44
Candida intermedia[1]	*n*-alkanes (C_{16}–C_{22})	0·81	0·96	0·35
Methylococcus sp.[4]	methane	1·01	1·34	0·29

[1]Johnson (1967a), [2]Pirt & Callow (1960), [3]Pirt (1957), [4]Harwood & Pirt (1972), [5]Harrison *et al.* (1972), [6]Takahashi *et al.* (1970)

for cell carbon, energy for growth and energy for maintenance. It can be seen that the overall yield based on carbon is rather constant varying from 0·90 to 1·34; hence this has some predictive value for growth yields.

Ultimately energy is provided in the form of ATP, the so-called energy currency of the cell, which is used to drive all the energy-requiring processes.

8.2 Assimilated carbon

When a substrate acts as both carbon and energy source for an organism accurate estimation of the part of the carbon source which is assimilated and that part which is dissimilated to provide energy is difficult. However, adequate estimations often can be made by one or more of the following four methods.

The first method is based on labelling the substrate carbon with ^{14}C and determining the amount of label which appears in the cell carbon. This method has been applied to *Streptococcus faecalis* by Bauchop & Elsden (1960) and to yeast by Kormancikova *et al.* (1969).

In the second method the substrate carbon assimilated is determined from the balance,

$$\begin{array}{ccc} \text{total substrate} = & \text{substrate utilized to} + & \text{substrate utilized to} \quad 8.1 \\ \text{utilized} & \text{provide cell carbon} & \text{provide energy} \\ (\Delta S) & (\Delta S_c) & (\Delta S_E) \end{array}$$

If we divide the substrate balance by Δx, that is the amount of biomass produced, we obtain

$$\Delta S/\Delta x = \Delta S_c/\Delta x + \Delta S_E/\Delta x \qquad 8.2$$

which can be written as

$$1/Y = 1/Y_c + 1/Y_E \qquad 8.3$$

If β is the fraction of the biomass which is carbon and γ is the fraction of the substrate which is carbon, it follows that

$$\beta Y_c = \gamma \qquad 8.4$$

Using Eqn *8.4* to substitute for Y_c, in Eqn *8.3* and rearranging we obtain

$$Y_E = \gamma Y/(\gamma - \beta Y) \qquad 8.5$$

This method was used by Stouthamer & Bettenhausen (1973).

The third method, applicable to aerobic processes, is to estimate Y_E from the stoichiometry of oxygen consumption and energy substrate dissimilation. It is assumed that the oxygen is used exclusively as the final electron acceptor in the energy-yielding process (Hernandez & Johnson, 1967). This assumption may not always be valid since, in some processes, oxygen is not simply reduced but is incorporated by an oxygenase into carbon compounds. However, in this case there is often a simple stoichiometry between carbon substrate utilized and oxygen incorporated, for example, one atom of oxygen to one molecule of *n*-paraffin, which enables the incorporated oxygen to be separated from that which is reduced (Harrison *et al.*, 1969).

A fourth method which could be used to measure the amount of substrate dissimilated to provide energy is to measure the amount of an end product of energy metabolism; examples are lactic acid from glucose in lactic fermentations, ethanol from glucose in yeast fermentations and acetic acid produced from ethanol by acetic acid bacteria. Comparisons of assimilated and dissimilated portions of carbon and energy sources are given in Table 8.3.

TABLE 8.3   Assimilated and dissimilated carbon and energy substrate

Organism	Conditions	Glucose carbon assimilated (%)	Glucose dissimilated to provide energy (%)
Streptococcus faecalis[1]	Rich medium, anaerobic	2	98
Saccharomyces cerevisiae[2]	Rich medium, anaerobic	2	98
Saccharomyces cerevisiae[2]	Rich medium, aerobic	10	90
Aerobacter cloacae[3]	Minimal medium, aerobic	55	45

[1]Bauchop & Elsden (1960), [2]Kormančičková *et al.* (1969), [3]Pirt (unpublished)

8.3 Maintenance energy

8.3.1 Definition

It has for long been postulated (Pirt, 1965) that microbes and cells require energy both for growth and for other so-called maintenance purposes. Certain specific maintenance functions recognized now are: turnover of cell materials, osmotic work to maintain concentration gradients between the cell and its exterior, and cell motility.

Maintenance energy is defined in the following way. The total energy source consumed (ΔS_E) is supposed to contribute a ration of energy for growth (ΔS_G) and a ration for maintenance (ΔS_M). Conceivably ATP may be generated and lost through its hydrolysis not being coupled to growth or maintenance functions; such lost ATP would normally be indistinguishable from maintenance energy. We can write

$$Y_E = \Delta x / \Delta S_E = \Delta x / (\Delta S_G + \Delta S_M) \qquad 8.6$$

where Δx is the amount of biomass formed. When the maintenance energy is zero, that is $\Delta S_M = 0$, we have the 'true' growth yield given by

$$Y_{EG} = \Delta x / \Delta S_G \qquad 8.7$$

This is the maximum possible value of the growth yield from the energy source. We suppose that with a given amount of biomass, x, energy source is consumed at a constant rate for maintenance given by $(ds/dt)_m = mx$ where m is a constant called the maintenance coefficient. The balance for energy source utilization is given by

$$\begin{array}{ccc} \text{total rate of} & = \text{rate of consumption} & + \text{rate of consumption} \\ \text{consumption} & \text{for growth} & \text{for maintenance} \end{array}$$

that is

$$\mu x / Y_E = \mu x / Y_{EG} + mx \qquad 8.8$$

hence

$$1 / Y_E = 1 / Y_{EG} + m / \mu \qquad 8.9$$

Alternatively we can write

$$q_E x = \mu x / Y_{EG} + mx \qquad 8.10$$

where q_E is the metabolic quotient for the energy source, hence

$$q_E = \mu / Y_{EG} + m \qquad 8.11$$

From Eqn 8.9 it follows that if m is constant the graph of $1/Y_E$ against $1/\mu$ (Fig. 8.1a) will be a straight line with slope, m and intercept, $1/Y_{EG}$ on the

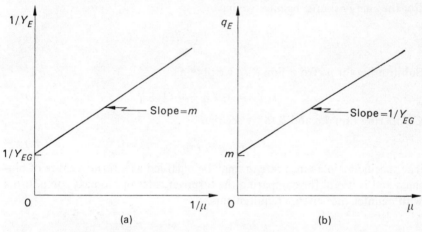

FIG. 8.1   Graphical methods for the calculation of the maintenance coefficient (m) and the maximum growth yield ($1/Y_{EG}$) from plots of Eqn 8.10 and 8.12.

ordinate. Alternatively m may be estimated from the plot of q_E against μ (Fig. 8.1b). When the specific growth rate has been varied by means of chemostat culture, a straight-line relation between $1/Y_E$ and $1/\mu$ has generally been observed (Pirt, 1965; Stouthamer & Betthausen, 1973). In the original plots of Eqn 8.9 (Pirt, 1965) the growth yield included the assimilated carbon source, that is the reciprocal growth yield was $1/Y_E + 1/Y_c$ (Eqn 8.3); however, this did not affect the calculation of m.

8.3.2 Maintenance as consumption of biomass

To account for the growth yield from the energy source decreasing with growth rate, Herbert (1958) and Marr et al. (1963) represented the maintenance energy requirement as consumption of biomass through endogenous metabolism, thus we have

net growth = total growth − biomass consumed endogenously

that is

$$dx = \mu_T x.dt - ax.dt \qquad\qquad 8.12$$

where μ_T is the total specific growth rate and a is called the specific maintenance rate. From Eqn 8.12 we derive

$$dx/dt = (\mu_T - a)x \qquad\qquad 8.13$$

The apparent specific growth rate is given by

$$\mu = \mu_T - a \qquad\qquad 8.14$$

For the energy source balance we have

$$\frac{\mu x.dt}{Y_E} = \frac{\mu_T x.dt}{Y_{EG}} \qquad\qquad 8.15$$

Substituting for μ_T from Eqn 8.15 we obtain

$$1/Y_E = 1/Y_{EG} + a/\mu Y_{EG} \qquad\qquad 8.16$$

Comparing Eqn 8.9 and 8.16 we see that

$$m = a/Y_{EG} \qquad\qquad 8.17$$

The specific maintenance rate, a, may be regarded as a turnover rate of biomass and is useful for comparing the turnover rates of biomass components with maintenance energy requirements.

8.3.3 Magnitude and control of maintenance energy

Table 8.4 sets out representative values of the maintenance energy coefficients of protists under various conditions. They range from 0·5 (mmol ATP.g dry biomass^{-1} h^{-1}) for the yeast without added sodium chloride to 220 (mmol ATP g dry biomass^{-1} h^{-1}) for *Azotobacter* fixing nitrogen at the high dissolved oxygen tension. Clearly the maintenance energy can be a massive part of the total energy consumed. Stouthamer & Bettenhausen (1973, p. 62) estimated that when *Klebsiella aerogenes* grows anaerobically with glucose as the energy source at a specific growth rate of 0·1 h^{-1}, maintenance energy accounts for 90% of the energy source consumed. The value of m for *K. aerogenes* found by Stouthamer & Bettenhausen was about 3 or 4 times that reported previously for the organism (Pirt, 1965); the difference could have been the result of the energy source being in excess in the culture referred to by Stouthamer whereas it was growth-limiting in the culture referred to by Pirt.

The results of Watson (1970) and of Stouthamer & Bettenhausen (1973), given in Table 8.4, show that the ionic strength of the medium has a big effect

TABLE 8.4   Some values of maintenance energy requirements of protists with glucose as the energy source

Organism	Special growth conditions	Maintenance energy	
		m (g energy source/ g dry biomass.h)	m_{ATP} (mmol ATP/ g dry biomass.h
Lactobacillus casei[1]		0·135†	1·5
Aerobacter cloacae[2]	aerobic, glucose-limited	0·094	14*
Klebsiella aerogenes[3]	anaerobic, tryptophan-limited NH$_4$Cl (2 g/l)	2·88	39
Klebsiella aerogenes[3]	anaerobic, tryptophan-limited NH$_4$Cl (4 g/l)	3·69	50
Saccharomyces cerevisiae[4]	anaerobic	0·036	0·52
Saccharomyces cerevisiae[4]	anaerobic NaCl (1·0 M)	0·360	2·2
Penicillium chrysogenum[5]	aerobic	0·022	3·2*
Azotobacter vinelandii[6]	Fixing nitrogen, dissolved oxygen tension (0·2 atm)	1·5*	220*
Azotobacter vinelandii[6]	Fixing nitrogen, dissolved oxygen tension (0·02 atm)	0·15*	22*

* Assuming 26 moles ATP produced/mole glucose, i.e. $P/O = 2$
† Assuming 1 mole glucose produces 2 moles ATP

[1]De Vries et al. (1970), [2]Pirt (1965), [3]Stouthamer & Bettenhausen (1973), [4]Watson (1970), [5]Righelato et al. (1968), [6]Nagai & Aiba (1972)

on the maintenance energy. Watson found that adding sodium chloride (1·0 M) to the medium increased the m_{ATP} fourfold (it also affected the metabolic pathways which decreased the amount of ATP produced from glucose), and Stouthamer & Bettenhausen found that increasing NH$_4$Cl concentration from 0·2% to 0·4% increased m_{ATP} by 11 (mmol ATP.g dry biomass^{-1} h^{-1}). These observations suggest that a large part of the maintenance energy ration is required to perform osmotic work to maintain concentration gradients between the cell and its exterior.

The effects of temperature and pH value on the maintenance energy requirement have not been investigated systematically. The fact that over the temperature range from 15 to 36°C the growth yield from glucose for Escherichia *coli* was constant (Harrison & Loveless, 1971a) suggests that m may be little affected by temperature. However, the same authors reported that pH affected the glucose yield substantially. Decrease in pH from 6·6 to 5·4 was accompanied by an increase in the q_{O_2} which would correspond to an increase in m_{ATP} of 16 (mmol ATP.g dry biomass^{-1} h^{-1}).

The extent to which turnover of cell material can use up energy may be estimated by comparison of known turnover rates with the specific maintenance rate. The rate of protein turnover in growing bacteria is estimated to be 0·006 h^{-1} (0·6%/h) though it rises to ten times that value when growth ceases (Mandelstam, 1960). Taking protein to constitute 60% of the biomass the turnover rate in growing cells corresponds to a biomass turnover rate of 0·6 × 0·006 h^{-1}, that is 0·0036 h^{-1}. In contrast, the specific maintenance rate for *E. coli* is about 0·06 h^{-1}; hence protein turnover appears to account for an insignificant part of the maintenance energy in growing cells. The protein turnover rate of *E. coli* in the non-growing state (6%/h) would correspond to a specific maintenance rate of 0·036 h^{-1}. Brooks & Meers (1973) have observed that, when the energy substrate is growth-limiting, intermittent feeding of the energy source decreases the Y_E value; this effect could be the result of increased protein turnover when growth stops. Another major cell constituent which turns over is RNA (Mandelstam, 1960) but the amount in growing cells (< 0·5%/h) accounts for only a minor part of the maintenance energy. In some bacteria turnover of cell wall material may account for a major part of the maintenance energy during growth (Mauck *et al.*, 1971).

Azotobacter species have often been cited for their exceptionally high q_{O_2} values. A number of workers have now shown that these exceptionally high values only occur when these bacteria are fixing nitrogen at high dissolved oxygen tensions, since the maintenance energy is proportional to the dissolved oxygen tension (Nagai & Aiba, 1972); this could be a means whereby the cell removes oxygen which could be an inhibitor of the highly reduced nitrogenase system.

It seems that maintenance functions may account for a major part of the energy requirement of protists. The great variability of maintenance energy and the factors which control it call for more systematic study.

8.4 Effects of maintenance energy

8.4.1 On relation of growth rate to energy substrate concentration

The requirement of cells for maintenance energy should affect the relation between growth rate and concentration of energy substrate. From Eqn *8.11* and *2.20* we have

$$q_E = \mu/Y_{EG} + m = q_m s/(s + K_s) \qquad\qquad 8.18$$

hence

$$\mu = \mu_m s/(s+K_s) - m Y_{EG} \qquad 8.19$$

where $\mu_m = q_m Y_{EG}$. When $s \gg K_s$

$$\mu = \mu_m - m Y_{EG} \qquad 8.20$$

Putting $s = s_m$ when $\mu = 0$ then

$$s_m = m K_s Y_{EG}/(\mu_m - m Y_{EG}) \qquad 8.21$$

and when $s = 0$

$$\mu = -m Y_{EG} \qquad 8.22$$

The graphical representation of Eqn 8.19 is given in Fig. 8.2. According to the

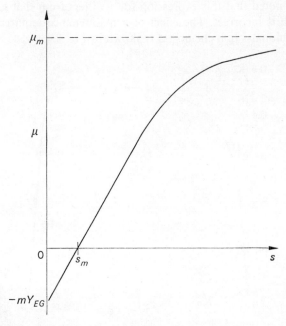

FIG. 8.2 Relation (Eqn 8.19) between specific growth rate (μ) and concentration of energy source (s) when there is a constant maintenance coefficient (m).

model, therefore, the maximum growth rate is decreased if there is a requirement for maintenance energy and the growth rate becomes zero at a finite concentration (s_m) of energy substrate. The deviation of Eqn 8.19 from Eqn 2.21 will be significant only if m is very large; *Azotobacter* fixing nitrogen at a high oxygen partial pressure could be a case in point.

8.4.2 On growth in a chemostat culture

Existence of the maintenance energy requirement calls for modification of the quantitative theory of the chemostat. For the growth rate we still have

$$dx/dt = (\mu - D)x \qquad\qquad 8.23$$

but, if the energy source is the growth-limiting substrate, then we have

$$ds/dt = D(s_r - s) - \mu x/Y_{EG} - mx \qquad\qquad 8.24$$

For μ we substitute from Eqn 8.19. In the steady state when $dx/dt = ds/dt = 0$, $\mu = D$

$$\tilde{s} = (D + mY_{EG})K_s/(\mu_m - mY_{EG} - D) \qquad\qquad 8.25$$

and

$$\tilde{x} = DY_{EG}(s_r - \tilde{s})/(D + mY_{EG}) \qquad\qquad 8.26$$

It should be noted that this expression for $\tilde{x}$ differs from that given by Pirt (1965), which is incorrect. The effect of a maintenance requirement on the steady-state biomass in a chemostat is shown in Fig. 8.3.

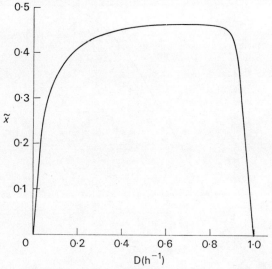

FIG. 8.3   Steady-state biomass concentration ($\tilde{x}$) as a function of dilution rate (D) in a chemostat when there is a maintenance requirement for the growth-limiting substrate: $\mu_m = 1 \cdot 0$ h^{-1}, $K_s = 0 \cdot 005$ g l^{-1}, $s_r = 1 \cdot 0$ g l^{-1}, $Y_{EG} = 0 \cdot 5$ g biomass (g substrate)$^{-1}$, $m = 0 \cdot 08$ g substrate (g.biomass.h)$^{-1}$.

8.4.3 On growth in a batch culture

The effect of a maintenance requirement for the growth-limiting substrate in a batch culture will be to decrease μ (predicted by Eqn 8.19) and the growth yield. The amount of energy source consumed during an interval, t, would be given by

$$\Delta s_E = (x - x_0)/Y_{EG} + m \int_{x_0}^{x} x \, dt \qquad 8.27$$

where x_0 and x are the biomasses at times 0 and t respectively. By determining Δs_E at two different growth rates and evaluating the integral either analytically or graphically, the value of m may be determined. Monod (1942) applied this method to determine m for *Escherichia coli* varying μ by limiting the aeration. He thus found that with lactate as the energy source m was zero. This failure to detect any maintenance energy may have been because the growth rate difference was too small, or because the ATP yield increased when the aeration was limited so as to mask the maintenance energy.

8.5 ATP yield (Y_{ATP})

Bauchop & Elsden (1960) defined the ATP yield of biomass as

$$Y_{ATP} = MY_E/n \qquad 8.28$$

where n = moles of ATP made available to the organism by the metabolism of one mole of energy source; M = molecular weight (g) of energy source and MY_E is the molar growth yield when Y_E is expressed as g dry biomass produced/g energy source. Bauchop & Elsden found that Y_{ATP} was about 10·5 and independent of the nature of the organism and the environment. This being so, it follows from Eqn 8.28 that the amount of energy source required to synthesize a given amount of biomass should be directly proportional to the number (n) of moles of ATP produced per mole of energy source. Factors which influence ATP production such as uncoupling of oxidative phosphorylation, or changes in metabolic pathways will influence Y_E. Also maintenance energy will affect Y_{ATP} through its effect on Y_E as shown by Eqn 8.19. Thus the maximum value of Y_{ATP}, that is, Y_{ATP}^{max}, is given by the following relation derived from Eqn 8.11,

$$q_{ATP} = \frac{\mu}{Y_{ATP}^{max}} + m_{ATP} = \frac{\mu}{Y_{ATP}} \qquad 8.29$$

in which the energy source unit is one gram molecule of ATP. Stouthamer & Bettenhausen (1973) pointed out that the effect of maintenance energy on Y_{ATP} has been greatly underestimated and is one factor responsible for the wide range of Y_{ATP} values found.

Theoretical calculations of Y_{ATP} based on known energy requirements for synthesis of bacterial cells from glucose, ammonia and inorganic salts give Y_{ATP} values up to 28·8 (Stouthamer, 1973). The theoretical value is little influenced by substituting amino acids for ammonia but addition of nucleic acid bases increases the theoretical maximum Y_{ATP} to 32·1 (Stouthamer, 1973). Experiments by Hernandez & Johnson (1967) suggest that in practice the Y_E value is little affected by adding amino acids, nucleic acid bases and vitamins to a minimal medium for bacteria. The theoretical maximum Y_{ATP} is much lower (4·85) for biomass synthesis from carbon dioxide (autotrophic metabolism), and if acetic acid is the energy source the theoretical maximum is 10·0 (Stouthamer, 1973). In their reappraisal of the Y_{ATP} data for cell synthesis from carbohydrate, Stouthamer & Bettenhausen (1973) conclude that Y_{ATP}^{max} values of 25 can be achieved with bacterial cultures. Also they estimate that the moles of ATP produced per atom of oxygen taken up (the P/O ratio) for some aerobic bacterial species is 1·9.

We can conclude that the net yield of ATP molecules (n) per mole of energy source is the basis on which biomass yields from the energy source can be predicted and that the value of Y_{ATP} can vary from about 5 to 32 g dry biomass per mole ATP depending on the nature of the carbon source. Also the value found can be decreased much below the maximum theoretical value through use of ATP for maintenance energy and by uncoupling of ATP production from energy source dissimilation.

8.6 Conditions affecting metabolic fates of carbon and energy substrates

The pathways, end products and ATP yields of carbon and energy source metabolism can be affected by dissolved oxygen concentration, pH value, temperature, ionic strength and trace element deficiency; these effects are discussed in the special sections on these factors. The carbon and energy source metabolism can also be affected by the specific growth rate and whether the carbon source is growth-limiting or in excess.

Specific growth rate affects the fate of glucose fermented by *Lactobacillus casei* (De Vries *et al.*, 1970). At low growth rates, acetate, formate and ethanol are the end products, whereas at high growth rates the glucose is fermented entirely to lactic acid. In the aerobic utilization of glucose by *Klebsiella aerogenes* (nitrogen-limited), Tempest *et al.* (1967) found that large amounts of α-ketoglutarate were formed at a very low growth rate but not at high growth rates.

When *K. aerogenes* is provided with excess glucose and oxygen the organisms accumulate large amounts of pyruvate, whereas when glucose limited, the organisms convert practically all of the carbon source to biomass and carbon dioxide (Harrison & Pirt, 1967). *Aerobacter cloacae* when provided with excess glucose and oxygen accumulates acetate (Pirt, 1957). It thus appears that, when bacteria are provided with excess of carbon and energy source and excess oxygen, incomplete oxidation may occur. Also, when energy source is supplied in excess, it may be stored as an energy reserve constituting a high proportion of the biomass dry weight (Wilkinson & Munro, 1967).

8.7 Utilization of mixed carbon sources

Frequently in batch cultures, different carbon sources in a mixture are utilized sequentially, for example glucose before lactose by *Escherichia* (Monod, 1942). Thus the utilization of one substrate may be inhibited by the other, an effect accounted for by repression of enzyme synthesis and inhibition of the permease (Miles & Pirt, 1973).

In chemostat cultures when a mixture of two carbon sources is supplied and the growth is carbon limited, both substrates may simultaneously be utilized over a wide range of dilution rates; examples are: glucose and lactose with *Klebsiella aerogenes* (Baidya *et al.*, 1967), glucose and maltose with *K. aerogenes* (Harte & Webb, 1967), glucose and acetate with *K. aerogenes* (Pirt, unpublished), glucose and xylose with *Escherichia coli* (Standing *et al.*, 1972). When the carbon source is not growth-limiting, for example with nitrogen-limited growth, then preferential utilization of the one carbon source occurs; for instance, acetate is used before glucose in a nitrogen-limited *Pseudomonas* species (Ng & Dawes, 1973) whereas both carbon sources are simultaneously utilized when the culture is carbon limited (Trilli & Pirt, unpublished).

8.8 Carbon dioxide supply

Carbon dioxide is required for two purposes: (*i*) it is the carbon source for autotrophic bacteria and algae; (*ii*) it is an essential intermediate in the metabolism of all organisms and consequently it must be present in all culture media.

The concentration of dissolved carbon dioxide will depend on the balance of production of carbon dioxide and its removal by gas transfer from the liquid to the gaseous phase. Thus we have for a batch culture

accumulation rate (R) = production rate − rate of transfer from liquid to gas

that is

$$R = q_{CO_2}x - K_L a(c - c_s) \qquad 8.30$$

where K_L is a constant, a = gas–liquid interfacial area per unit volume, c = actual concentration of dissolved carbon dioxide, c_s = concentration of dissolved carbon dioxide in equilibrium with gas phase (compare oxygen transfer, Section **9.6.1**). Putting H for Henry's constant and p_l and p_g for the partial pressures of carbon dioxide in liquid and gaseous phases respectively Eqn *8.30* becomes

$$R = q_{CO_2}x - K_LaH(p_l - p_g) \qquad\qquad 8.31$$

If there is no production of carbon dioxide, that is $q_{CO_2} = 0$, then it is necessary to add carbon dioxide to the gas phase and we should have $p_g > p_l$. For a chemostat culture we need to add to the right-hand side of Eqn *8.30* a term, $-Dc$, to represent the washout of dissolved carbon dioxide, however, often this term can be neglected, then for the steady state when $R = 0$ we derive from Eqn *8.31*.

$$\tilde{p}_l = \tilde{p}_g + q_{CO_2}\tilde{x}/K_LaH \qquad\qquad 8.32$$

The automatic control of dissolved carbon dioxide by means of a carbon dioxide electrode has been achieved by Ishizaki *et al.* (1973).

8.9　Carbon dioxide–carbonate equilibria in solution

In solution carbon dioxide forms carbonic acid which dissociates to give bicarbonate and carbonate ions, thus

$$CO_2 + H_2O \rightleftharpoons H_2CO_3$$
$$H_2CO_3 \rightleftharpoons HCO_3^- + H^+$$
$$HCO_3^- \rightleftharpoons H^+ + CO_3^{2-}$$

The total concentration of dissolved carbon dioxide S is taken to include the carbonic acid, thus

$$S = [CO_2] + [H_2CO_3] = HP_{CO_2} \qquad\qquad 8.33$$

where the square brackets denote concentrations and P_{CO_2} is the partial pressure of the dissolved carbon dioxide. The value of H for an aqueous solution at 30°C is 0·030 mol/atm, a value about thirty times that for oxygen. For the ionization constants of H_2CO_3 we have

$$K_1 = \frac{[H^+][HCO_3^-]}{S} = 10^{-6\cdot3} \quad \text{at } 30°C \qquad\qquad 8.34$$

hence

$$\log [HCO_3] = \log S + pH - pK_1 \qquad\qquad 8.35$$

where $pK_1 = -\log K_1$. For the ionization of HCO_3^- we have

$$K_2 = \frac{[H^+][CO_3^{2-}]}{[HCO_3^-]} = 10^{-10\cdot3} \quad \text{at } 30°C \qquad\qquad 8.36$$

hence

$$\log [CO_3^{2-}] = \log [HCO]_3^- + pH - pK_2 \qquad 8.37$$

Strictly the concentrations should be replaced by activities, however, for the present purpose it is assumed that the activity coefficients are unity. The calculated concentrations of bicarbonate and carbonate ions at various pH values with P_{CO_2} at 10^{-2} atm are given in Table 8.5. At pH 10 it seems unlikely that the P_{CO_2} could be maintained at 10^{-2} atm in a culture because of the high associated bicarbonate concentration.

TABLE 8.5 The concentrations of bicarbonate and carbonate ions at various pH values with carbon dioxide partial pressure at 10^{-2} atm

pH Value	$[HCO_3^-]$ (M)	$[CO_3^{2-}]$ (M)
4	$1 \cdot 50 \times 10^{-6}$	$7 \cdot 5 \times 10^{-10}$
7	$1 \cdot 50 \times 10^{-3}$	$7 \cdot 5 \times 10^{-7}$
10	$1 \cdot 50$	$7 \cdot 5 \times 10^{-4}$

Temperature, 30°C; the concentrations were calculated from Eqn 8.35 and 8.37 with $S = 0 \cdot 03 \times 10^{-2}$ M

8.10 Influence of carbon dioxide partial pressure on growth and metabolism

Most studies on the effects of partial pressure of carbon dioxide on growth have been limited to comparing the effects of presence or absence of carbon dioxide. The results of Dagley & Hinshelwood (1938) show that the relation between bacterial growth rate and P_{CO_2} is of hyperbolic form with the half-maximum growth rate for *Klebsiella aerogenes* occurring at a P_{CO_2} of 3×10^{-4} atm, that is the partial pressure of carbon dioxide in air at 1 atm. From the solubility of carbon dioxide given by S in Eqn 8.33, it follows that $K_s = 0 \cdot 9 \times 10^{-5}$ M, which is about the same as the value for carbon substrates generally. The maximum growth rate of *K. aerogenes* required a P_{CO_2} of about 10^{-3} atm. The results of Lwoff & Monod (1947) showed that the growth rate of *Escherichia coli* in a glucose minimal medium is substantially decreased with $P_{CO_2} = 3 \times 10^{-5}$ atm, however, this effect could be counteracted by the addition, at 10^{-4} M, of glutarate, succinate, asparagine or glutamate, the last being the most effective. The effects of the supplements were not additive. Growth virtually ceased with $P_{CO_2} = 6 \times 10^{-6}$ atm and could not be restored by the supplements. It seems that the supplements used cannot substitute for carbon dioxide but act as carbon dioxide sparing agents.

It has been suggested that P_{CO_2} can be critical for microbial functions other than growth (Wimpenny, 1969) but, in general, the evidence for these effects is not convincing. In contrast Ishizaki *et al.* (1973), by means of automatic control of dissolved carbon dioxide concentrations, found that the over-production of inosine by *Bacillus subtilis* is inhibited strongly by $P_{CO_2} > 0.03$ atm, moreover, the results indicate that carbon dioxide rather than the bi-carbonate ion was the inhibitory agent.

8.11 Hydrocarbons as carbon and energy sources

Hydrocarbons, because of their abundance and low cost have become of major importance as substrates in fermentations to produce single cell protein and for transformation into other more valuable products. A salient feature of hydrocarbons as substrates is their insolubility, which probably accounts in part for the neglect of hydrocarbons in microbiology. The solubility of *n*-alkanes reaches a peak of about 60 mg/l with chain lengths from C_2 to C_4 and then decreases with chain length according to the expression

$$\log H = 4.526 - 0.588n \qquad \qquad 8.38$$

where H is the solubility in mg/l and n is the number of carbon atoms in the molecule (Johnson, 1964). Table 8.6 shows that the solubility falls below

TABLE 8.6   Solubility of *n*-alkanes in water at 25°C (Data from Johnson, 1964)

n-Alkane	Concentration of saturated solution (M)
Hexane	1.1×10^{-4}
Heptane	2.6×10^{-5}
Octane	5.8×10^{-6}
Decane	3.1×10^{-7}
Dodecane	1.7×10^{-8}
Tetradecane	9.8×10^{-10}

10^{-5} M at octane. Since, in general, K_s values for carbon and energy sources are $> 10^{-5}$ M (Table 2.1) it seems improbable with the higher homologues that uptake of dissolved hydrocarbon can be fast enough to account for the specific growth rates, which exceed 0.2 h^{-1}. Johnson (1964) has calculated from diffusion considerations that 'even with a Michaelis constant of zero, appreci-able growth rates cannot be obtained on dodecane and higher hydrocarbons unless a special mechanism for substrate uptake is available'. This special mechanism appears to be uptake of the hydrocarbon directly from the oil phase (Humphrey & Erickson, 1972). Hence oil droplet size and oil–water interfacial area will be important factors influencing the growth rate. Essentially, with the higher alkane homologues, the growth rate eventually becomes limited by the oil–water surface area and this is reflected by the onset of linear growth in a batch culture. Models for the chemostat culture of microbes on insoluble oils taking into account direct transfer of hydrocarbon from oil to cell have been formulated (Humphrey & Erickson, 1972). From the generalized representation of the expected results with different dilution rates in a chemostat (Fig. 8.4) Humphrey & Erickson draw the following conclusions.

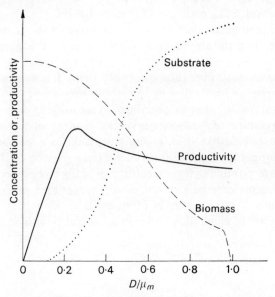

FIG. 8.4 Generalized representation of effects of dilution rate on biomass concentration, substrate (hydrocarbon) concentration and productivity, that is biomass output rate: D=dilution rate, μ_m=maximum specific growth rate (redrawn from Humphrey & Erickson, 1972) for growth on hydrocarbons.

'(1) Dispersed phase continuous culture systems generally give poor substrate utilization rates except at low dilution rates. (2) Cell concentrations decrease rather sharply with dilution rates. [This has been confirmed by the results of Munk *et al.* (1973).] (3) Optimum productivities ($D\tilde{x}$) do not occur at near washout conditions. Rather they occur at relatively low values of D/μ_{max}, generally in the range of 0·2 to 0·4.'

Comparison of the growth yields on hydrocarbons (Table 8.2) shows that on carbon conversion to biomass the yield is about the same as it is with other substrates, however, a major difference between growth on hydrocarbons and growth on carbohydrates is the much greater requirement for oxygen with hydrocarbons (Section **9.2**). The growth yield from a hydrocarbon was much increased by using a 'fed batch' (Section **21.1**) with carbon-limited growth instead of simple batch culture with hydrocarbon in excess (Yoshida *et al.*, 1973).

8.12 Dispersion of hydrocarbons in aqueous media

A convenient means of adding solid hydrocarbons to a culture is to use another inert hydrocarbon as a solvent; thus 2, 6, 10, 14-tetramethylpentadecane was used to dissolve C_{20} *n*-alkanes (Johnson, 1964).

Surfactants, acting as emulsifying agents, serve to increase the oil–water interfacial area in a culture; recommended agents for this purpose are polyoxyethylene(20) oleyl ether (Kobayashi, *et al.*, 1967) and polyoxyethylene-polypropylene monostearate (Someya *et al.*, 1970). It is evident also that the yeasts and bacteria which utilize hydrocarbons themselves synthesize emulsification agents. A slimy viscous material with emulsifying properties has been isolated from cultures of *Pseudomonas* species growing on decane (Maclennan & Pirt, 1970) and Suzuki *et al.* (1969) have isolated a trehalose lipid with emulsifying properties from cultures of bacteria growing on hydrocarbons. Goma *et al.* (1973) found that yeast cultures growing on hexadecane produced an agent which preferentially emulsified C_{16} rather than other alkanes. The chemical nature of these naturally produced emulsifiers is still obscure but they probably are of much significance in microbial growth on hydrocarbons.

CHAPTER 9

Oxygen demand and supply

9.1 Introduction

Study of oxygen demand and supply for microbial cultures began with Pasteur's discovery that yeast could take up oxygen and use it for the oxidation of carbon and energy sources. No significant progress in understanding the quantitative requirements for oxygen came until the 1920's when manometric respirometer techniques were developed to study respiration rates and the stoichiometry of substrate oxidations; however, the technique is convenient only for non-growing or 'resting' cells. In the 1930's the polarographic technique with an oxygen electrode was applied in respirometry (Baumberger, 1939), but some decades had to pass before the oxygen electrode was developed sufficiently to be used widely in studies on oxygen effects. In the 1930's oxidation–reduction or redox potential received much attention as a possible controlling factor in oxidation of energy sources, but the real significance of the redox potential of a culture has remained obscure. A new assessment of the significance of the redox potential in cultures is given in Section **9.5**.

In the 1940's the problem of providing sufficient oxygen for dense submerged cultures became important, largely through the need for industrial production of antibiotics. A similar requirement in the production of bakers' yeast, it was found empirically, could be overcome by 'incremental' feeds of sugar. In effect, this method controls the oxygen demand of the culture. The automatic control of dissolved oxygen concentration in submerged cultures is now possible, and application of this technique, particularly in steady-state chemostat cultures, is essential to elucidation of oxygen requirements.

9.2 Oxygen demand

In aerobic metabolism oxygen acts as the ultimate electron or hydrogen acceptor. This process is mediated by an oxidase enzyme. Besides its role as electron acceptor oxygen may, by means of an oxygenase enzyme, be incorporated into carbon substrates in their catabolism. The oxygenase function is

81

found notably in the catabolism of hydrocarbons and aromatic ring compounds.

A first estimate of the total amount of oxygen required in energy metabolism may be obtained from the stoichiometry of the oxidation. For example, for glucose oxidation during growth of *Escherichia coli* we may write $C_6H_{12}O_6 + 6O_2 \rightarrow 6CO_2 + 6H_2O$. The amount of carbon source oxidized is equated with that required for energy supply and from this amount the growth yield from the energy source (Y_E) is derived (Section **8.2**).

When hydrocarbons are used as the carbon source then oxygen is required not only to produce energy but also for raising the oxidation level of the substrate up to that of biomass. It follows from calculation of the oxygen balances that production of biomass from a hydrocarbon requires much more oxygen than production from carbohydrate (Darlington, 1964). The amount of oxygen required may be calculated as follows (Johnson, 1964). Let $A=$ amount of oxygen required for the combustion of 1 g of substrate to CO_2, H_2O (and NH_3 if the substrate contains nitrogen); $B=$ amount of oxygen required for combustion of 1 g dry biomass to CO_2, H_2O and NH_3; $Y=$ growth yield in g dry biomass/g substrate; $C=$ amount of oxygen used during the production of 1 g dry biomass.

We have the balance

oxygen required to pro-	=	oxygen required to	−	oxygen required
duce 1 g dry biomass		burn substrate con-		to burn 1 g dry
		sumed to produce		biomass
		1 g dry biomass		

that is

$$C = A/Y - B \qquad\qquad 9.1$$

The value of B is estimated to be 41·7 mmol oxygen/g dry biomass; $A=$ 33·33 mmol oxygen/g glucose or 107·1 mmol oxygen/g alkane (CH_2). Now with $Y=0·5$ for glucose and 0·81 for an alkane, the amount of oxygen required for growth on alkanes is found to be 3·6 times as much as that for growth on glucose. A comparison of growth yields based on oxygen is given in Table 8.2. It follows that if oxygen transfer rate (Section **9.6.1**) limits the output of the fermenter, the maximum output rate with glucose as substrate would be 3·6 times that with *n*-alkanes.

The theoretical maximum growth yield for oxygen when glucose is the carbon source may be estimated as follows. Assume that 2ATP moles are produced per atom of oxygen, that is P/O $= 2$, and that 2ATP are produced per mole of glucose during glycolysis, then the total ATP produced per mole of glucose oxidized will be 26. Taking Y_{ATP} to be 25 g dry biomass/mol ATP

(Section **8.5**) then the theoretical maximum growth yield for oxygen ($Y_{x/0}$) is 3·39 g dry biomass/g oxygen. The experimental values of $Y_{x/0}$ for bacteria and fungi with carbohydrate as energy source are generally about one. The large difference between the theoretical maximum and the observed value of $Y_{x/0}$ is attributed to maintenance energy and loss of ATP production.

In oxidation reactions which involve oxygenases as well as oxidases the two enzymes must compete for the oxygen. The two enzyme activities may be separately evaluated when the stoichiometry of the oxygenase reaction is known. Thus Harrison *et al.* (1969) considered that in the oxidation of decane by a *Pseudomonas* species the oxygen uptake rate of the energy-yielding process ($q_{O_2}^E$) could be represented by

$$q_{O_2}^E = q_{O_2}^T - q_{\text{decane}}$$

where $q_{O_2}^T$ is the total metabolic quotient for oxygen and q_{decane} the quotient for decane uptake, the quotients being expressed on a molar basis. Here it is assumed that the oxygenase reaction involves equimolar amounts of decane and oxygen, hence the q_{decane} should equal the oxygen uptake rate mediated by the oxygenase.

9.3 Oxygen solubility

The solubility of oxygen in aqueous media is a few mg/l with air at 1 atm pressure (see Table 9.1). This is slight compared with the amount of oxygen

TABLE 9.1   Oxygen solubility in water and Henry's constant (H)

Temperature (°C)	Oxygen concentration* (mg/l)	H (mg/l atm)
25	8·10	38·8
35	6·99	33·4

* In equilibrium with air at 1 atm $P_{O_2} = 0·209$ atm);
calculated from Eqn *9.3*

which can be consumed in a culture. As much as 50 g oxygen/l may be required by very dense cultures, consequently, it is necessary to renew the dissolved oxygen by transfer of oxygen from gas to liquid medium.

The main factors which affect the solubility of oxygen are partial pressure of oxygen, temperature and other solutes in the medium. The effect of oxygen pressure in the gas phase (p_g) is expressed by Henry's law which states that

$$c_s = Hp_g \qquad\qquad 9.2$$

where c_s = saturation concentration of oxygen, and H is Henry's constant. The partial pressure is also called the *oxygen tension*.

Oxygen solubility decreases with increase in temperature. The effect in water was determined experimentally by Truesdale *et al.* (1955) and represented by the empirical relation

$$c_s = 14\cdot16 - 0\cdot3943T + 0\cdot007714T^2 - 0\cdot0000646T^3 \qquad 9.3$$

where c_s is the saturation concentration of oxygen (mg/l) in pure water at $T°C$ with air at 1 atm (air contains 20·9% oxygen, oxygen tension = 158·8 mmHg). The standard error of the curve is 0·05 mg/l.

The activity of oxygen in a saturated solution is given by the relation $a = fc_s$ where f is the activity coefficient. We can equate the activity of the gas with its partial pressure putting $a = p_g$, so that Henry's constant (H) is the reciprocal of the activity coefficient. Dissolved salts generally decrease the solubility of oxygen, which means that the activity is increased. This effect can be large, for instance, in saturated sodium chloride solution the solubility is approximately halved, that is, the activity coefficient is doubled. Truesdale *et al.* (1955) give the effects of sea water salinities on the solubility of oxygen.

9.4 Measurement of amount of dissolved oxygen

9.4.1 Measurement of oxygen concentration

Two basic methods are available for the determination of dissolved oxygen concentration. One method is titration, usually by the Winkler method (Lunge & Keane, 1908), which depends upon oxidation of manganous salt to permanganate by oxygen, then titrating iodine liberated from iodide by the permanganate. The method is subject to interference by organic matter, and consequently its applicability to culture media is limited. The second method depends on gasometry (Roughton & Scholander, 1953). In this method oxygen is swept out of the solution by the generation of carbon dioxide, then the gas formed is analysed. This method, unlike the titrimetric method, is not subject to interference by organic matter. Neither method, however, is rapid or sensitive enough to be applicable in the range of dissolved oxygen concentrations which normally limit respiration rates of microbes.

9.4.2 Measurement of dissolved oxygen tension (DOT)

Oxygen tension is measured electrometrically by the oxygen electrode, the current from which is directly proportional to the oxygen tension. Thus the oxygen electrode measures the oxygen activity of a solution and not the concentration, this is shown, for instance, by saturating water with sodium chloride which halves the oxygen concentration but does not affect the oxygen electrode reading with a given gas pressure. It can be seen that conversion of the oxygen electrode reading into oxygen concentration, involves the assumption of some value of H. However, for most purposes, it is convenient to express the oxygen electrode readings in terms of dissolved oxygen tension (DOT). The oxygen electrode has been used successfully to measure and control DOT down to a fraction of a mmHg (Maclennan & Pirt, 1966).

Dissolved oxygen tension is also measured by immersing a tube of oxygen permeable material, such as Teflon, in the culture, then purging the tube with nitrogen or other oxygen-free gas. The dissolved oxygen diffuses into the tube, and the DOT is related to the amount of oxygen in the gas stream determined by some form of gas analyser (Phillips & Johnson, 1961).

9.5 Redox potential

9.5.1 Definition

Oxidation–reduction or redox potential measures the tendency of a solution to give or take up electrons. The potential is measured by immersing an inert electrode, usually of platinum, in the solution and observing the difference in potential between the platinum and a reference electrode.

A reversible oxidation–reduction process can be generally formulated as

$$a(\text{oxidant}) + b\text{H}^+ + ne \rightleftharpoons c(\text{reductant})$$

where a and c are the number of moles of oxidant and reductant respectively and n is the number of electrons transferred in the process. The redox potential (E_h) of the process is given by

$$E_h = E_0 + \frac{RT}{nF} \ln \frac{[\text{oxidant}]^a [\text{H}^+]^b}{[\text{reductant}]^c} \qquad 9.4$$

where the quantities in square brackets are the activities of the reactants concerned (concentration may replace activity when the activity coefficient is unity), R is the gas constant and F is the faraday quantity of electricity. The hydrogen electrode reaction is given by

$$2\text{H}^+ + 2e \rightleftharpoons \text{H}_2$$

for which

$$E_h = E_0 + \frac{RT}{2F} \ln \frac{[\text{H}^+]^2}{P_{\text{H}_2}} \qquad 9.5$$

where P_{H_2} is the partial pressure of the hydrogen gas. The *standard hydrogen electrode* is obtained when $[\text{H}^+] = 1 \cdot 0$ M and $P_{\text{H}_2} = 1$ atm, then $E_h = E_0$, which potential is defined as zero. If the potential is not measured against a standard hydrogen electrode but rather against a calomel electrode or some other reference electrode, then, if E_1 is the potential measured and E_h^r is the potential of the reference electrode, $E_h = E_1 + E_h^r$. Values of E_h^r for various reference electrodes are given by Jacob (1970, p. 101).

Equation 9.4 shows that E_h is pH dependent. If we rewrite Eqn 9.4 in the form

$$E_h = E_0 + 2 \cdot 30 \cdot \frac{RT}{nF} \log \frac{[\text{oxidant}]^a}{[\text{reductant}]^b} - 2 \cdot 30 \cdot \frac{bRT}{nF} \text{pH} \qquad 9.6$$

then the slope of the graph of E_h against pH value, when oxidant and reductant activities are constant, will be $-2 \cdot 30 \cdot bRT/nF$, which represents the change in E_h for a pH change of 1 unit. If $b/n = 1$ this value of $\Delta E_h/\Delta \text{pH}$ is 58 mV at 30°C.

The electrometric measurement of E_h is discussed in detail by Jacob (1970). The E_h reading is highly dependent on the preparation of the platinum electrode, for instance the degree of polishing of the electrode (Jacob, 1970, p. 100).

Indicators, which may be used to measure E_h value, are listed by Jacob (1970). The E_h range of the indicator may be considered with reference to methylene blue for which we write

$$MB^+ + H^+ + 2e \rightleftharpoons MBH$$

and

$$E_h = E_0 + 2 \cdot 30 \cdot \frac{RT}{2F} \log \frac{[MB^+]}{[MBH]} - 2 \cdot 30 \cdot \frac{RT}{2F} \cdot \text{pH} \qquad 9.7$$

that is, in mV,

$$E_h = E_0 + 29 \cdot \log \frac{[MB^+]}{[MBH]} - 29 \, \text{pH} \qquad 9.8$$

at 30°C. The value of E_0 is the E_h when $[MB^+] = [MBH]$ and pH = 0. At pH 7 we write $E_h = E_0^7 = E_0 - 203$ mV. The indicator range may be taken to be the E_h range in which the amount of the oxidized form of the indicator changes from 90% to 10% of the whole, that is log $[MB^+]/[MBH]$ changes by 2. This means that the E_h changes by about 60 mV and we can write the E_h range of methylene blue as E_0 or $E_0^7 \pm 30$ mV.

9.5.2 Redox potential of an oxygen solution

The significance of the redox potential of a culture medium is usually obscure because the controlling electron donor and acceptor have not been identified. However, it has been shown that oxygen can control the E_h value of complex culture media (Barnes & Ingram, 1956). The E_h of an oxygen solution may be deduced from consideration of the reaction

$$O_2 + 2H^+ + 4e \rightleftharpoons 2OH^-$$

for which we have

$$E_h = E_0 + \frac{RT}{4F} \ln \frac{Po_2[H^+]^2}{[OH^-]^2} \qquad 9.9$$

where Po_2 is the dissolved oxygen tension and it is assumed that the reaction is in equilibrium. Substituting for $[OH^-]$ from the dissociation constant for water, that is, $K_w = [OH^-][H^+]$, changing to logarithms to base 10, and rearranging Eqn 9.9, we obtain

$$E_h = E_0' + \frac{2 \cdot 30 \cdot RT}{4F} \log Po_2 + 2 \cdot 30 \cdot \frac{RT}{F} \log [H^+] \qquad 9.10$$

The value of E_0' is 1229 mV (Clark 1960, p. 79) so that Eqn 9.10 becomes

$$E_h = 1229 + 14 \cdot 8 \log Po_2 - 59 \text{ pH} \qquad 9.11$$

where E_h is measured in mV and Po_2 in atmospheres. Experimental results show that the relation between E_h and $\log Po_2$ is rectilinear in accordance with Eqn 9.11 (Jacob, 1970), however, the E_0' value is much lower than the theoretical value. With air at 1 atm ($Po_2 = 0 \cdot 209$ atm) and pH = 7, the theoretical E_h value is 806 mV whereas the maximum value found is about 400 mV (Jacob, 1970, p. 119). Thus the E_h value of an oxygen solution is anomalous and must be determined empirically. Also, the slope of the graph of E_h against $\log Po_2$ is empirically found to be 60 mV/log Po_2, a value which holds down to 10^{-9} atm oxygen (Schuldiner et al., 1966), whereas according to Eqn 9.11 the slope should be 15 mV per log Po_2. These discrepancies between the theoretical and experimental values may occur because the oxidation–reduction processes in an oxygen solution are not in equilibrium, as the theory assumes.

A number of workers have proposed that E_h may be used as a measure of dissolved oxygen tension in culture media (Jacob, 1970, p. 116). However, at the oxygen tensions required in aerobic cultures, E_h measurements, because of their logarithmic response, are much less sensitive to DOT than are measurements with the membrane-covered oxygen electrode, the current from which has a linear response to DOT. In contrast, the E_h may be a useful alternative to DOT measurement when log DOT is about three logarithms below that of the saturation value with air at 1 atm.

9.6 Oxygen transfer from gas to liquid to biomass

9.6.1 Gas to liquid transfer

Despite considerable advance in knowledge of the factors governing the transfer of oxygen from gas to liquid, it often still limits the rate of aerobic microbial processes. The principles of gas absorption developed for oxygen transfer are applicable also to the absorption of other gaseous substrates such as methane, carbon dioxide and nitrogen in cultures.

The simplest concept of a gas absorption process is the *stationary liquid film theory*. At the interface between the gas and the liquid phases there is supposed to be a stationary liquid film in which there is a concentration gradient of dissolved gas (Fig. 9.1a). At the interface proper, the liquid is assumed to be

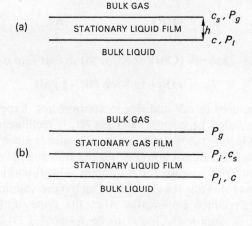

FIG. 9.1   Oxygen concentration gradients in gas to liquid transfer and liquid to bio-mass transfer; c and P refer to the dissolved oxygen concentrations and oxygen tensions respectively. (a) Liquid film gradient in gas to liquid transfer. (b) Two-film model of gas to liquid transfer. (c) Liquid film gradient in liquid to biomass transfer.

saturated with the gas which will be at concentration c_s; in the bulk of the liquid the gas concentration is c. If we consider unit volume of liquid which has an interfacial area, a, then the rate of absorption of the gas per unit volume will be

$$R_s = \frac{dc}{dt} = \frac{k_L a}{h}(c_s - c) \qquad\qquad 9.12$$

where h is the thickness of the stationary film and k_L is a constant dependent on the diffusion coefficient of the gas. Because, in general, the value of h is not known, it is usual to combine it with k_L and write

$$dc/dt = K_L a(c_s - c) \qquad\qquad 9.13$$

Equation 9.13 is the normal relation for the diffusion of a substance across a barrier, where $(c_s - c)$ represents the driving force. On the analogy of electric current flow, where the current = the potential difference/resistance, the factor $K_L a$ in Eqn 9.13 can be seen as the inverse of a resistance to oxygen transfer. The oxygen absorption rate can be expressed in terms of oxygen partial pressures in gas and liquid (p_g and p_l respectively) by making the substitutions, $c_s = H p_g$ and $c = H p_l$,

$$dc/dt = K_L a H (p_g - p_l) \qquad 9.14$$

The efficiency of oxygen absorption is normally expressed in terms of either $K_L a$, or $K_L a H$. The maximum oxygen absorption rate with a given oxygen partial pressure in the gas phase occurs when $c = 0$; then we have

$$(dc/dt)_{max} = K_L a c_s = K_L a H p_g \qquad 9.15$$

There are various elaborations on the stationary film theory (Aiba et al., 1965). One of these is the *two-film theory* according to which, besides the stationary liquid film there is supposed to be a stationary gas film at the interface (Fig. 9.1b). The oxygen absorption rate in the steady state can then be represented as

$$dc/dt = K_g a H (p_g - p_i) = K_L a H (p_i - p_l) = K_t a H (p_g - p_l) \qquad 9.16$$

where $1/K_G$ is the gas film resistance, $1/K_L$ is the liquid film resistance and $1/K_t$ is the overall resistance given by

$$1/K_t = 1/K_G + 1/K_L \qquad 9.17$$

In systems with a high degree of turbulence the resistance of the gas film is negligible (Aiba et al., 1965, p. 135; Finn, 1954) and then $K_t = K_L$.

In summary, it follows from Eqn 9.14 that the gas to liquid oxygen transfer rate is proportional to the interfacial area, the oxygen partial pressure and Henry's constant and varies inversely as the liquid film resistance ($1/K_L$).

9.6.2 Liquid to biomass transfer

In the transfer of gas through a liquid to the biomass surface an additional resistance to diffusion is involved, namely that of the stationary liquid film surrounding the biomass (Fig. 9.1c). Finn (1954) deduced from data on the respiration of yeast cells suspended in a stagnant liquid medium that the maximum concentration difference in the liquid film surrounding a yeast cell, that is $(c - c_b)$ in Fig. 9.1(c), would be 2×10^{-7} M. In an agitated suspension the stationary film thickness and consequently the concentration difference across it would be smaller. This concentration difference is considered negligible compared with the DOT value of about 10^{-6} to 10^{-5} M (Table 2.1) at which respiration rate starts to be oxygen limited (Borkowski & Johnson, 1967).

Tsao (1970) has suggested that microbial cells may penetrate the stationary liquid film at a gas–liquid interface and even concentrate in it owing to the surface activity of the cells. This penetration of the liquid by the cells should, in effect, decrease the film thickness and consequently increase $K_L a$. Tests of this theory remain to be made.

9.7 Measurement of $K_L a$ value in absence of biomass

9.7.1 With oxygen-absorbing reagents

Oxygen absorption by solutions of chemical reagents such as pyrogallol (Dixon & Elliott, 1930) and sodium sulphite with cupric ions as catalyst (Cooper et al., 1944) may be used as measures of oxygen solution rate. The principle of such methods is that the rate of oxidation of the reagent is limited only by the rate of oxygen transfer from gas to liquid. The copper-catalysed sulphite oxidation has been used widely to determine oxygen solution rates in fermenters. In this method the rate of oxygen absorption is independent of sulphite concentration above about 0·01 M (Phillips & Johnson, 1959). The rate of oxygen absorption is dependent on the nature of the catalyst since the rate can be increased several fold by substituting cobaltous ions for cupric ions (Pirt & Gillett, 1955). This effect of the nature of the catalyst may be caused by occurrence of the chemical reaction not only in the bulk of the solution but also in the stationary film so that increase in the reaction rate, in effect, decreases the apparent thickness of the liquid film. During sulphite oxidation the dissolved oxygen tension is practically zero so that the oxygen absorption rate is given by $K_L a H p_g$.

Because of differences between the processes in the liquid films, the liquid film resistances in a sulphite solution and in a microbial culture are not necessarily identical. Comparison of maximum oxygen solution rates in a bacterial culture and in copper-catalysed sulphite oxidation (Pirt & Callow, 1958a) showed that $K_L a$ values for sulphite oxidation were higher by a factor of 1·3 to 2·0 than those for a culture. On the other hand, Phillips & Johnson (1961) found that $K_L a$ values in both bacterial and filamentous mould cultures could be 50% greater than in copper-catalysed sulphite oxidation. These differences could result from variations in either K_L or in a.

Hsieh et al. (1969) measured oxygen solution rates by means of the glucose oxidase system. In this system, three enzymes are used to oxidize glucose to gluconic acid and water. The oxygen absorption rate is measured by titration of the acid produced. From knowledge of the apparent Michaelis constant of the enzyme system and the maximum velocity of the oxidation, the dissolved oxygen concentration (c) can be calculated and $K_L a$ is calculated by means of Eqn 9.13. The values of $K_L a$ thus determined were reported to be 'somewhat higher' than those found by copper-catalysed sulphite oxidation.

9.7.2 With an oxygen electrode

With a fast-responding oxygen electrode the oxygen solution rate may be determined by direct measurement of the rate of increase in dissolved oxygen tension after it has been lowered by passing oxygen-free gas through the system. By substituting $dc/dt = H dp_l/dt$ in Eqn 9.14 we obtain

$$dp_l/dt = K_L a(p_g - p_l) \qquad 9.18$$

which on integration gives

$$\ln(p_g - p_l) = \ln(p_g - p_{l0}) - K_L a t \qquad 9.19$$

where $p_{l0} = $ DOT value when $t = 0$. The graph of $\ln(p_g - p_l)$ against t will have a slope of $-K_L a$. This method has advantages over the chemical methods in that it can be carried out in many different media and does not involve a chemical reaction which could affect the liquid film resistance. However, good agreement between the copper-catalysed sulphite oxidation and the oxygen electrode method have been reported (Chain & Gualandi, 1954).

9.8 Measurement of $K_L a$ value during a culture

9.8.1 By steady-state kinetics

The value of $K_L a$ under culture conditions can be measured readily by means of a chemostat culture in a steady state. The oxygen balance in the culture is given by

net accumulation	=	amount in medium added	+	transfer from gas to liquid	−	uptake by biomass	−	wash out

that is

$$dc/dt = DHp_g + K_L a H(p_g - p_l) - x q_{O_2} - DHp_l \qquad 9.20$$

where $D = $ dilution rate, $x = $ biomass per unit volume and it is assumed that the medium added is saturated with oxygen. For most cultures, the first term on the right-hand side of Eqn 9.20 is negligible compared with the second; also the fourth term is negligible compared with the third, hence we can write

$$dc/dt = K_L a H(p_g - p_l) - x q_{O_2} \qquad 9.21$$

In the steady state, $dc/dt = 0$ then

$$p_l = p_g - xq_{O_2}/K_LaH \qquad 9.22$$

The value of the oxygen uptake rate (xq_{O_2}) is determined either by gas analysis or by the non-steady-state method (Section **9.8.2**). With a fixed dilution rate, xq_{O_2} can be maintained constant over a wide range of p_g values and a plot of p_l against p_g will have the slope $-1/K_LaH$ until oxygen begins to limit the respiration rate. Such a straight line relation was obtained in a culture of *Klebsiella aerogenes* down to a DOT value less than 15 mmHg (Harrison & Pirt, 1967).

9.8.2 By non-steady state kinetics

In a batch culture the oxygen uptake rate per unit volume of culture is given by

$$dc/dt = K_La(c_s - c) - q_{O_2}x \qquad 9.23$$

where x is the biomass at time t. By stopping the oxygen supply the first term on the right hand side of Eqn *9.23* becomes zero so that $dc/dt = q_{O_2}x$. Thus, from the slope of the graph of c against t (Fig. 9.2) the oxygen uptake rate

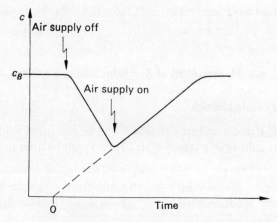

FIG. 9.2   Changes in dissolved oxygen concentration (c) used to calculate K_La by unsteady-state kinetics (Section **9.8.2**).

$q_{O_2}x$ can be calculated. When the air supply is recommenced the dissolved oxygen concentration increases according to Eqn *9.23* and the slope of the graph gives the value of dc/dt. By rearrangement of Eqn *9.23* we obtain

$$c = \frac{-1}{K_La}\left(\frac{dc}{dt} + q_{O_2}x\right) + c_s \qquad 9.24$$

From this relation we see that, with constant $K_L a$, a plot of c against $(dc/dt + q_{O_2}x)$ will have a slope of $-1/K_L a$. The term $(dc/dt + q_{O_2}x)$ may be varied either by altering the biomass value or the partial pressure of oxygen in the gas phase. The interval during which the changes in c are followed must be short enough to have only a negligible effect on the biomass concentration. This method was applied by Bandyopadhyay et al. (1967). Fujio et al. (1973) preferred the integrated form of Eqn 9.23 for the evaluation of $K_L a$, that is

$$\ln(1 - c/c_B) = -K_L a t \qquad\qquad 9.25$$

where $t = 0$ when $c = 0$ and c_B is the value of c just before the air supply is stopped (see Fig. 9.2), which assumes that $dc/dt \approx 0$. By substituting $c = Hp_l$ and $c_B = Hp_{lB}$ where p_l and p_{lB} are dissolved oxygen tensions, Eqn 9.25 becomes

$$\ln(1 - p_l/p_{lB}) = -K_L a t \qquad\qquad 9.26$$

Then $K_L a$ can be determined from the slope of the graph of $\ln(1 - p_l/p_B)$ against t and there is no need to know the value of Henry's constant as is the case with Eqn 9.24.

9.8.3 With oxygen-limited growth

When respiration is oxygen-limited we can put, in Eqn 9.23, $c \approx 0$ and $dc/dt = 0$, hence

$$K_L a c_s = K_L a H p_g = q_{O_2}x \qquad\qquad 9.27$$

The oxygen uptake rate $(q_{O_2}x)$ is determined by gas analysis and thus $K_L a H$ or $K_L a$ may be calculated.

CHAPTER 10

Aeration and agitation methods

10.1 Introduction

The simplest means for the aeration of microbial cultures is by growth on the surface of a static liquid or solid medium. This method is commonly used in the laboratory and in the older aerobic industrial fermentations for production of vinegar and citric acid. Originally penicillin was produced by surface culture of the *Penicillium* mould, however, the need for penicillin stimulated study of the submerged culture of aerobic organisms and soon this led to the development of the basic principles of culture aeration.

Kluyver & Perquin (1933) developed the shake-flask technique primarily for the submerged culture of moulds for which the natural habitat is the surface of a substrate. The development of the shake-flask technique was a very important advance because it enabled aerobic organisms to be grown to high density under homogeneous conditions, and consequently simplified study of the physiology of the organisms.

The laboratory stirred fermenter has evolved into the most versatile and fully controlled means for submerged culture of both protists and animal and plant tissue cells. Rolling fermenters (for example, Ugolini *et al.*, 1959) have been proposed as an alternative to stirred fermenters as a means of avoiding the shear of stirrer blades, which it has been claimed can damage the organism. Although shear of filamentous organisms obviously occurs on some occasions, there is no definitive evidence that shear is detrimental to the function of a culture. Even mammalian cells, which may be regarded as the most delicate of structures, have been grown successfully in laboratory fermenters with high rates of stirring (Klein *et al.*, 1971). Perhaps shear of biomass is not a prime cause of injury but only enhances the effects of an adverse growth medium. Where shear is proved to be detrimental, the roller fermenter or vibration may be a suitable alternative means of agitation. Vibrating agitators (Ulrick & Moore, 1965) have the advantage that they do not require a gland, which is commonly required to seal a stirrer shaft in the fermenter and may lead to contamination if the seal fails.

94

10.2 Agitation and mixing

Agitation of cultures has two functions: it assists mass transfer between the different phases (gas, liquid and solid) present in the culture; and it mixes the culture and so maintains homogeneous chemical and physical conditions in the culture—this is particularly important to disperse the biomass, for heat transfer and for mixing a continuous feed of substrate.

Mixing efficiency can be measured by the time it takes to disperse a small volume of a dye solution added to the liquid in the fermenter (Aiba *et al.*, 1965, p. 173). In fermenters, good mixing becomes increasingly difficult to achieve with increase in the size of the culture vessel, increase in viscosity, and —if there is a continuous feed of substrate—decrease in the flow rate of the substrate solution. In a study of the latter case, Hansford & Humphrey (1966) found that decreased mixing efficiency in a 5 l chemostat decreased the growth yield of yeast in a glucose-limited culture. The effect became marked with a dilution rate $< 0.03 \, h^{-1}$. The mixing was found to be improved by having a multiple substrate feed in place of a single point feed.

Agitation affects mass transfer from gas to liquid in three basic ways. Firstly, it disperses the gas into smaller bubbles and thus increases the interfacial area. Secondly, it increases the gas–liquid contact time because circulation of the gas bubbles in eddies causes gas 'hold up' which may be defined as follows. Let $V_0 =$ volume of ungassed liquid, $V =$ volume of liquid–gas dispersion, then the hold up volume

$$V_H = V - V_0 \qquad\qquad 10.1$$

The mean contact time of the gas is V_H/F where F is the gas flow rate. Increase in gas hold up increases the interfacial area per unit volume. Also change in the hold up volume will alter the extent to which oxygen is taken up from the gas. Thirdly, the increased turbulence caused by agitation will decrease the thickness of the stationary liquid film and consequently increase the value of K_L (Section 9.6.1).

10.3 Baffled, vortex and airlift systems of aeration and agitation

Agitation in a stirred fermenter can be either free or baffled (Fig. 10.1). The insertion of the baffle prevents vortex formation and increases turbulence in the fluid. In order to aerate in the baffled system air must be introduced by a sparger below the liquid level. In the vortex system air is entrained by the vortex and dispersed in the liquid by the impeller. There is intense turbulence in the region of the impeller with vortex agitation but outside the impeller zone the liquid is circulated with little turbulence.

Virtually full turbulence can be obtained by means of four equally spaced baffles of width 0·08 to 0·1 times the vessel diameter and running the full depth of the liquid (Finn, 1954).

The draught tube (Fig. 10.1c), used in the Waldhof fermenter developed for

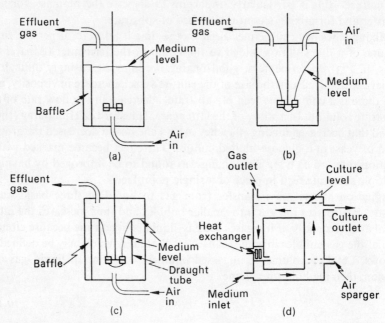

FIG. 10.1   Diagrammatic representations of agitation-aeration systems in stirred fermenters: (a) baffled system; (b) vortex system; (c) baffled system with draught tube; (d) I.C.I. 'pressure cycle' fermenter in which the culture circulates round the vessel in the direction of the arrows (Gow *et al.*, 1973).

yeast production in Germany during World War II, creates a vortex despite the baffle. Chain *et al.* (1952, p. 94) found that the Waldhof fermenter behaves like a plain vortex system on the 5 l scale. It would be interesting to see whether this conclusion remains valid on scale-up of the process.

Imperial Chemical Industries, in their development of single cell protein production by growth of bacteria on methanol, have introduced the 'air lift' principle for aeration and mixing in a large scale fermenter (Fig. 10.1d). This system has the advantage of greater mechanical simplicity than a large stirred fermenter. The $K_L a$ value attainable in this system can be calculated from Eqn 9.22 for the methanol process if we assume that $Y_{x/0} \approx 0.44$ (Table 8.2), $\mu = 0.2\ h^{-1}$, $x = 20\ g/l$, c_s with air supplied at 3 atm is $18 \times 10^{-3}\ g/l$ and the mean $c = 9 \times 10^{-3}\ g/l$. Then $K_L a$ is about $1000\ h^{-1}$, which is of the order attained with stirred fermenters.

10.4 Impeller design

The most efficient type of impeller is the vaned disc with vertical vanes depicted in Fig. 10.2. The optimum diameter is about 0·4 times the vessel dia-

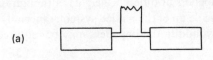

(a)

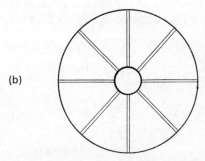

(b)

FIG. 10.2   Vaned disc impeller. (a) Elevation of impeller as fitted for baffled system with main depth of blade below the disc. When used in the vortex system the impeller is fitted in the reverse way; (b) plan showing eight vanes.

meter. The vertical vanes cause centrifugal flow of the liquid, in contrast to the axial flow which results with inclined blades. Sulphite oxidation rates showed that in a vortex system with a stirrer at 1000 rev/min the oxygen solution rates were several times higher with vertical vanes than with inclined vanes (Chain et al., 1952, p. 88); however at 2000 rev/min the difference was slight. Hence at normal stirrer speeds in the vortex system (< 1500 rev/min) the vertical vanes are optimal.

At a given stirrer speed the oxygen solution rate obtained with the vaned disc is the same whether it is situated at half the medium depth or just above the base of the fermenter; however, in the vortex system, with the higher impeller position less power is consumed (Chain et al., 1952, p. 84). In the author's laboratory it has been found that if the impeller is situated close to the base of a laboratory fermenter with vortex aeration, the disc becomes superfluous, possibly because the base of the fermenter acts like the disc.

In the vortex system, the oxygen solution rate obtained with eight vanes on the disc was several fold greater than that with four vanes and there was no difference between 8 and 16 vanes (Chain et al., 1952, p. 90), hence the optimum number of vanes is taken to be eight.

The optimum vane height is about $\frac{1}{6}$ of the impeller diameter (Chain *et al.*, 1952, p. 89; Finn, 1954).

The vortex system requires only one impeller. In a laboratory scale baffled system Chain *et al.* (1952, p. 82) found no increase in sulphite oxidation rate on adding more impellers at different heights, although the power consumption increased. On scale up of the baffled system, especially in media with non-Newtonian viscosity, additional impellers may be required.

10.5 Effect of stirrer rate

The effects of rate of stirring on the oxygen solution rates in the vortex and baffled aeration systems are depicted in Fig. 10.3. In the baffled system $K_L a$ is

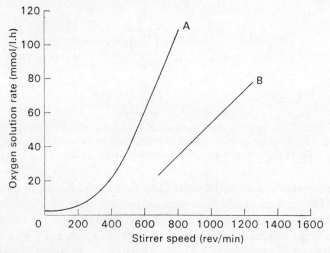

FIG. 10.3   Effect of stirrer rate on oxygen solution rate in baffled and vortex aeration systems with the same vaned disc impeller in a laboratory fermenter (based on data from Chain & Gualandi, 1954). Line A, baffled system; line B, vortex system.

roughly porportional to N^3 where N is the stirrer speed in rev/min (Chain & Gualandi, 1954; Finn, 1954). In the vortex system there is a high threshold rate of stirring before the oxygen solution rate begins to increase, but thereafter it increases linearly with stirrer rate.

10.6 Effect of air sparging

The effects of air sparging at different rates in the baffled and vortex systems are illustrated in Fig. 10.4 and 10.5. Figure 10.4 shows that the oxygen solu-

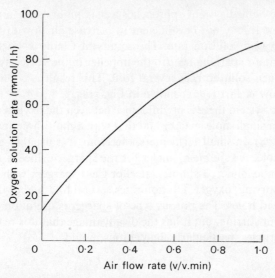

FIG. 10.4   Effect of air sparging on oxygen solution rate in a baffled stirred fermenter; v/v.min = volumes of air/volume liquid × minute (Chain *et al.*, 1952).

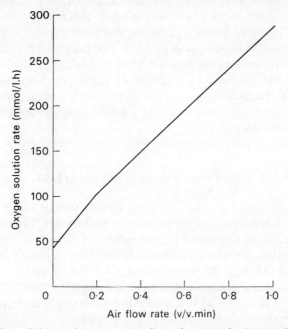

FIG. 10.5   Effect of air sparging on oxygen absorption rate of sulphite solution (copper catalyst) in vortex system (redrawn from Pirt & Callow, 1961). Single orifice sparger; vaned disc impeller, diameter 102 mm in vessel of 228 mm diameter; liquid volume, 12 l.

tion rate in the baffled system approaches a maximum as the air flow rate is increased, hence it may not be sufficient to increase air flow rate in order to increase the oxygen solution rate. The results of Chain & Gualandi (1954, p. 48) show that air sparging near to the impeller in the vortex system can increase the oxygen solution rate several fold. This result is confirmed by the results of Callow & Pirt (1961) shown in Fig. 10.5.

In the baffled system there is no difference between the oxygen solution rate obtained with a single-hole sparger and that with a multi-hole sparger (Chain *et al.*, 1952, p. 78). In small fermenters, therefore, it is usual to use a sparger with a single hole. A suitable diameter for the sparger orifice is 1 to 3 mm.

Porous ceramic, sintered stainless steel or glass spargers with micro holes give, without stirring, oxygen solution rates several fold higher than spargers with macro-sized holes. The porous type of sparger is useful on occasion for aeration without stirring, but it has the disadvantage that it is readily clogged by biomass particles, particularly filamentous moulds.

10.7 Effects of temperature and viscosity

For oxygen transfer in aqueous media the value of $K_L a \propto T^{1/2}$ where T is the absolute temperature (Aiba *et al.*, 1965, p. 151). This means that about 30°C the $K_L a$ value increases about 13% per 10° rise in temperature.

O'Connor (quoted by Aiba *et al.*, 1965, p. 151) suggested that $K_L a \propto (1/\eta)^{1/2}$ where η is the viscosity of the medium. Hattori *et al.*, 1972 studied the effect of increasing the viscosity by addition of carboxymethylcellulose to the medium. Their results indicate that $K_L a \propto \eta^{-0.35}$ for $\eta < 50$ cP and $K_L a \propto \eta^{-1.1}$ for $\eta > 50$ cP; thus 50 cP was a critical viscosity at which the effect became much more pronounced.

10.8 Effects of surface active agents and hydrocarbons

Culture media and culture products often include surface active agents and foam formation is evidence of the presence of such agents. The effects of surface active agents on $K_L a$ values are unpredictable. Aiba *et al.* (1965, p. 153) have systematized the observations available. They reported that sodium lauryl sulphate (10 mg/l) decreases both K_L and $K_L a$ by about 50%. With a larger amount of the detergent, $K_L a$ increased slightly whereas K_L remained constant, an effect attributed to decrease in bubble size. The decrease in $K_L a$ with the lower detergent concentration is unexplained.

Chain & Gualandi (1954, p. 42) reported that foam formation by undefined agents in a vortex aeration system decreased oxygen solution rate to $\frac{1}{10}$ of its normal value. Breaking the foam with an antifoam agent restored the oxygen solution rate to its normal value. It seems probable that foam formation in the vortex system results in circulation of the foam exhausted of oxygen and the entry of fresh air is excluded by the foam. Antifoam agents have been reported to decrease $K_L a$ apparently through increase in bubble size (Phillips & Johnson, 1961).

Addition of kerosene to a shake-flask has been reported to increase dramatically the $K_L a$ value (Mimura et al., 1969). Large amounts of kerosene— $> 10\%$ of the aqueous medium volume—were required to produce the effect, which is unexplained. However, oxygen is about 100-fold more soluble in oil than in water, which could endow the kerosene with an oxygen carrier property.

10.9 Effect of biomass

The presence of mould mycelium in the medium can severely decrease the $K_L a$ value. Penicillium mycelium at 20 g dry weight/l decreased the $K_L a$ value 40 to 50% (Chain & Gualandi, 1954, p. 43); Aspergillus mycelium at 20 g dry weight/l decreased $K_L a$ by 90% (Brierley & Steel, 1959). The latter authors also used paper pulp to simulate the mycelium effect. The high non-Newtonian viscosity caused by mycelium may largely account for its effect on the $K_L a$ value.

10.10 Power requirement

The $K_L a$ value is directly proportional to the power input per unit volume of culture (Finn 1954; Chain & Gualandi, 1954). For the fully turbulent system with a vaned disc turbine we can write

$$K_L a H \approx \alpha P / V \qquad\qquad 10.2$$

where P = power input, V = liquid volume, H = Henry's constant and α is a constant termed the power ratio by Chain & Gualandi (1954). These authors found that α had its maximum value of 138 mmol oxygen/h. watt. atm oxygen in the 5 l vortex system, however, the power ratio fell dramatically with increase in the size of the vortex system, becoming about 0·138 in a 2000 l fermenter. On the other hand, in the fully baffled system $\alpha \approx 100$ and it could be maintained in the liquid volume range from 5 to 10 000 l. It follows that the vortex aeration system is highly efficient on the small scale but, unlike the baffled system, is not amenable to scale up.

10.11 Foam formation and its control

Stable foam formation is an undesirable feature of the aeration of culture media containing organic matter. The foam stabilizers present in complex media or produced by the organisms are undefined and the control of foam formation is purely empirical. The detrimental effects of foam formation are as follows: (*i*) liquid culture can be transported by the foam into the air outlets of the culture; (*ii*) it may decrease the $K_L a$ value (see Section **10.8**); (*iii*) in chemostat cultures it varies the gas hold up and consequently makes the liquid volume uncontrollable.

Foam formation is normally prevented by an antifoam agent. Mechanical foam breakers have been recommended but their efficiency in practice has not been established definitely, a matter of importance in view of the additional mechanical complexity involved. Quantitative studies on the amounts of antifoam agents required to inhibit foaming have been reported by Ghosh & Pirt (1954) and by Pirt & Callow (1958b). Many antifoam agents, often of undefined composition, have been used. Polypropylene glycol (molecular weight, 2000) is particularly effective and biologically inert. However, despite antifoam agents, foam formation can on occasion prove uncontrollable except by decreasing the air flow through the culture.

10.12 Agitation-aeration systems for laboratory fermenters

In laboratory stirred fermenters ranging from about 0·1 to 10 l capacity, agitation and aeration by the vortex system is more convenient than the baffled system because it eliminates the sparger and baffles. Higher $K_L a$ values may be obtained in the vortex system by inserting a single hole sparger either underneath or beside the impeller. The elimination of baffles is especially useful in long continuous fermentations since accretions of biomass tend to form on the baffles. Another advantage of the vortex system is that it produces less foam than the baffled system. Baffled fermenters are more appropriate for the larger pilot plant scale of fermenter when vortex aeration becomes less efficient. For the culture of filamentous moulds in fully instrumented fermenters, especially for chemostat operation, there is a minimum possible size of fermenter, which is about 1·5 l in a vessel of 15 cm diameter, otherwise accretion of mould between instrument probes makes maintenance of culture homogeneity virtually impossible.

10.13 Shake-flask aeration

The shake-flask is generally the simplest vessel for submerged agitated batch culture. This technique, first developed by Kluyver & Perquin (1933), has since been extended to the culture of all types of protist and plant and animal tissue cells.

Rotary shakers (for example, Paladino, 1954) are generally preferred to the reciprocating shaker (for example, Kantorowicz, 1951) because rotary shaking avoids the splashing which throws the organism out of the liquid and causes it to adhere to the walls above the liquid level, an advantage particularly marked with filamentous mould cultures. Siliconing the walls of the flask is often effective in preventing accretion of biomass on the walls.

Both reciprocating and rotary shakers normally have throws in the range 25 to 50 mm, that is, the maximum distance moved by a point on the flask. Reciprocation is usually about 100 cycles/min and in rotary shaking the range is about 150 to 300 rev/min.

10.14 Factors influencing oxygen solution rates in shake-flasks

10.14.1 Effect of flask shape

Chain & Gualandi (1954, p. 33) found that with rotary shaking there is no significant difference between oxygen solution rates in round-bottomed and in conical flasks of the same capacity and with the same liquid volume. On the other hand, they found that with reciprocating shaking the oxygen solution rate in conical flasks is several fold greater than that in round-bottomed flasks.

The insertion of a baffle in a rotary-shaken flask can increase the oxygen solution rate ten or more fold (Chain & Gualandi, 1954, p. 37). Effective baffling is provided by four equally spaced indentations a few millimetres deep round the lower half of the flask; also a stainless steel coil is effective and has the advantage of being removable (Jensen et al., 1963). Another important role of the baffle is to improve the dispersion of immiscible substrates such as hydrocarbons. A troublesome feature of baffled shaking is that it causes splashing which tends to make filamentous moulds stick on the walls of the flask above the liquid medium.

10.14.2 Effect of liquid volume

With either rotary or reciprocating shaking the oxygen solution rate falls rapidly as the liquid volume is increased. Some results of Smith & Johnson (1954) depicted in Fig. 10.6 illustrate the effect.

10.14.3 Effect of shaker speed and throw

Oxygen solution rates in shake-flasks increase more than linearly with increase in either shaking speed or throw. With a liquid volume to flask volume ratio of 1:10, Chain & Gualandi (1954, p. 34) found that increasing the rotary shaking speed from 150 to 300 rev/min with a throw of 50 mm increased the oxygen solution rate 2·5-fold. Decreasing the throw from 50 to 32 mm decreased the oxygen solution rate by a factor of 0·5.

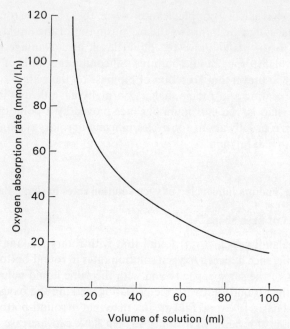

FIG. 10.6 Effect of liquid volume on oxygen absorption rate of sulphite solution (copper catalyst) in a shake flask from data of Smith & Johnson (1954): flask capacity, 500 ml; rotary shaking at 250 rev/min; throw, 50 mm.

10.14.4 Effect of biomass

Chain & Gualandi (1954, p. 35) found that with increases in dry weight of *Penicillium* mycelium up to 20 g/l the maximum oxygen solution rate decreased almost linearly, the total decrease being about 90%.Thus the presence of biomass can markedly decrease the oxygen absorption rate in a shake-flask.

10.14.5 Rate of gas diffusion through a cotton plug

Shake-flasks are usually plugged with cotton, which will form a barrier to diffusion of gas into or out of the flask. Schultz (1964) showed that the rate of oxygen diffusion into the flask (R_d, mmol/h) is given by

$$R_d = \frac{3600}{22 \cdot 4} \frac{D_a A}{L} (p_o - p_i) \qquad\qquad 10.3$$

where p_i and p_o are the partial pressures (atm) of oxygen inside and outside the flask respectively, A = cross-sectional area (cm^2) of cotton plug, L = length (cm) of plug, D_a = apparent diffusion constant (cm^2/sec). From experimental measurement of the oxygen diffusion rate with an oxygen electrode Schultz found that with the normal cotton plug of density, 0·05 to 0·08 g/cm^3,

$D_a \approx 0.19$ cm^2/sec, which is about 90% of the free diffusion rate. Hence the cotton plug imposes very little resistance to diffusion. The generalized equation for the oxygen solution rate in a shake-flask (R_s, mmol/h) derived by Schultz is

$$R_s = \left[\frac{22.4\,L}{3600\,D_a A} + \frac{1}{K_L a H V_l}\right]^{-1}(p_o - p_l) \qquad 10.4$$

where $K_L a$ = gas to liquid transfer coefficient (h^{-1}), H = Henry's constant for oxygen (mmol ml^{-1} atm^{-1}), V_l = volume of liquid (ml), p_l = dissolved oxygen tension (atm) in the liquid.

Application of Eqn 10.3 to a conical flask with a plug of 31 mm diameter and 35 mm long shows that the maximum rate of diffusion of oxygen into the flask would be 19 mmol/h with $p_l = 0$ and with air outside the flask at 1 atm. If the flask contained 25 ml of liquid then diffusion limitation of the oxygen solution rate would not occur until the rate reached 760 mmol/l.h, which is a large value compared with the usual demand rates. However, Schultz points out that as flask size, and consequently culture volume, is increased, the ratio A/L tends to stay constant and R_d is little affected so that diffusion through the plug could limit the oxygen solution rate in larger flasks.

10.15 Aeration in roller tubes

Tube rollers for culture agitation consist of racks of tubes rotated about an axis parallel to the tubes, generally at low speeds (about 60 rev/min) so that the centrifugal force is low. Hall (1960) evaluated the aeration capability of roller agitation and found the maximum oxygen solution rate to be about 15 mmol/l.h and virtually independent of the rotation rate in the range 6 to 60 rev/min. Hence the oxygen solution is much less than the value possible with rotary or reciprocating shaking. The advantage of tube rollers is that they are mechanically simpler and less expensive than shakers. Roller tubes are convenient for cultures of animal cells or protists in which oxygen demands are low.

10.16 Aeration in static submerged cultures

In a static submerged culture such as a test-tube culture of bacteria, or a monolayer culture of animal cells under a still medium the gas–liquid interfacial area is measurable. If we assume that the bulk liquid is homogeneous then, according to the stationary liquid film theory, the oxygen solution rate (mmol/min) is given by

$$R = K_L a(c_s - c) \qquad 10.5$$

where a is the area of the interface between liquid and gas; Schultz & Gaden (1956) found the value of K_L to be $1 \cdot 2 \times 10^{-3}$ cm/min for the diffusion of oxygen across a static surface during sulphite oxidation catalysed by copper ions. Hence the maximum oxygen solution rate $(K_L a c_s)$ in a test tube of diameter 16 mm will be 0·036 mmol/h or 3·6 mmol/l.h if the test tube contains 10 ml of liquid and $c_s = 0 \cdot 25$ mM; the experimental value of 1·8 mmol/l.h found by Smith & Johnson (1954) is of the same order. With such a low $K_L a$ value (about 10 h^{-1}) the static test tube culture can readily become oxygen limited. Diffusion of oxygen through a cotton plug into the tube, it can be shown (Section **10.14.5**), is unlikely to limit oxygen supply in a test tube culture.

CHAPTER 11

Effects of oxygen on microbial cultures

11.1 Introduction

Studies on the effects of oxygen on microbial cultures were initiated by Pasteur who discovered that yeast obtained energy for growth either by oxidizing sugar with oxygen or by anaerobic dissimilation of sugar to ethanol and carbon dioxide. Oxygen inhibits the anaerobic process and simultaneously decreases the glucose dissimilation rate. This inhibitory effect of oxygen, which has become known as the *Pasteur effect*, is an example of metabolic regulation by a substrate; however, the mechanism of the effect is still unknown.

The problem of how much oxygen is required for growth and other microbial functions is discussed in Section **9.2**; the present chapter is concerned with the responses of the organism to different dissolved oxygen tensions.

11.2 Oxygen-limited growth

In a batch culture with constant $K_L a$ value, when growth is oxygen-limited, the biomass growth rate (dx/dt) will become constant and the specific growth rate will decrease. In order to investigate the effects of oxygen limitation of growth it is best to use a chemostat culture under steady-state conditions so that the variation in specific growth rate can be avoided. The effect of oxygen limitation on biomass concentration in a chemostat culture can be predicted from Eqn 9.20. In the equation we substitute $c_s = Hp_g$, $c = Hp_l$, $q_{O_2} = D/Y_0$, where Y_0 is the growth yield from oxygen, and $dc/dt = 0$ for the steady state; then the biomass concentration will be

$$\tilde{x} = Y_0 \left(\frac{K_L a}{D} + 1 \right) (c_s - \tilde{c}) \qquad 11.1$$

In many systems $K_L a/D \gg 1$ then Eqn *11.1* reduces to

$$\tilde{x} = \frac{Y_0 K_L a}{D} (c_s - \tilde{c}) \qquad 11.2$$

The value of the steady-state oxygen concentration is given by $\tilde{c} = K_s D/$

($\mu_m - D$) where K_s is the saturation constant for oxygen. Figure 11.1 shows the predicted response of biomass concentration in a chemostat culture with oxygen limitation; with decrease in the dilution rate eventually the oxygen limitation disappears. This type of response was found by Pirt (1957) for oxygen-limited growth of *Klebsiella aerogenes*.

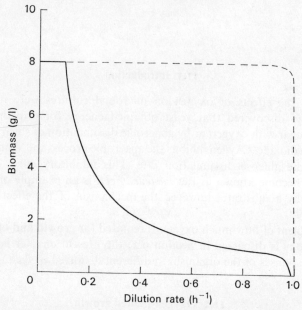

FIG. 11.1 Comparison of biomass production with oxygen-limited growth and carbon and energy substrate-limited growth in steady states of a chemostat culture. Continuous line, oxygen-limited biomass; broken line, carbon-limited biomass. Values of parameters used in Eqn *11.2* and Eqn *5.9*: $K_La = 100$ h^{-1}, $c_s = 8 \times 10^{-3}$ g/l, K_s(oxygen) $= 8 \times 10^{-5}$ g/l, $Y_0 = 1\cdot0$, $\mu_m = 1\cdot0$ h^{-1}, for carbon-limited growth $Y = 0\cdot5$, s_r for carbon substrate $= 16\cdot0$ g/l, K_s (carbon substrate) $= 0\cdot01$ g/l; maintenance energy is assumed to be nil.

11.3 Effect of dissolved oxygen tension on the oxygen uptake rate of growing biomass

The relation of the respiration rate of biomass to dissolved oxygen concentration is usually of the Michaelis–Menten type, that is

$$q_{O_2} = q_{O_2}^{max} \, c/(c + K_s) \qquad 11.3$$

If we substitute $c = Hp_l$ where p_l is the dissolved oxygen tension (DOT) and H is Henry's constant, also $K_s = HK_p$, then we obtain

$$q_{O_2} = q_{O_2}^{max} \, p_l/(p_l + K_p) \qquad 11.4$$

where K_p is the saturation constant measured in pressure units. Further, if μ is the specific growth rate and we assume the growth yield (Y_0) from oxygen is constant, we can substitute $q_{O_2} = \mu/Y_0$ and $q_{O_2}^{max} = \mu_m/Y_0$ and obtain

$$\mu = \mu_m p_l/(p_l + K_p) \qquad 11.5$$

which gives the relation between μ and p_l when oxygen is the growth-limiting substrate. The relation of c to μ for *Candida utilis* with acetate as the carbon source was found by Johnson (1967b) to conform with the Michaelis–Menten relation with $K_s = 1\cdot3 \times 10^{-6}$ M, that is $K_p = 0\cdot91$ mmHg. Often it is considered that respiration rate becomes independent of dissolved oxygen concentration above a certain value termed c_{crit}, which is defined in Fig. 11.2a. The value of

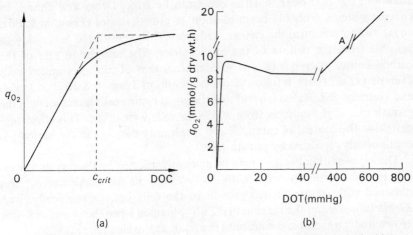

(a) (b)

FIG. 11.2 Effects of dissolved oxygen concentration on respiration rate (q_{O_2}): (a) relation between q_{O_2} and dissolved oxygen concentration (DOC) and definition of c_{crit}; (b) the response shown by a chemostat culture of *Pseudomonas* species ($\mu = 0\cdot1$ h^{-1}) with methanol as carbon source. About 'A' the q_{O_2} oscillated (from data of Maclennan *et al.*, 1971)

c_{crit} or the corresponding oxygen tension, p_{crit} will depend on the specific growth rate. For suspensions of unicellular organisms p_{crit} values up to 10 mmHg have been reported (Harrison *et al.*, 1969).

The influence of hyperbaric oxygen pressure, that is pressure > 0·21 atm or 159 mmHg, on respiration rate was investigated by Maclennan *et al.* (1971) using a pseudomonad growing on methanol. Their results Fig. 11.2b show that from DOT 30 to about 400 mmHg the respiration rate was constant, but at about 500 mmHg the q_{O_2} oscillated showing that there was instability in metabolic regulation. Above 560 mmHg the q_{O_2} dramatically increased with DOT. Since the biomass yield ($Y_{x/o}$) was maximal below 100 mmHg it seems that at DOT values > 100 mmHg and especially at values > 560 mmHg the maintenance energy was increased, possibly to overcome oxygen toxicity. Maclennan *et al.* (1971) point out that there is probably continuous variation in cell mechanisms over the whole DOT range, a point which would be missed by those studies confined to comparison of the effects of presence and absence of oxygen.

11.4 Influence of growth conditions on respiration rate of resting cells

Measurement of the respiration rate of resting, that is non-growing cells, has been a valuable parameter of metabolism. The extent to which the growth conditions can affect the respiration rate of the resting cells has not received much attention, but it is clear that the effect can be large. Decay and change in enzyme systems probably begin as soon as growth ceases (Fensom & Pirt, 1972) and to minimize the change, cells should be transferred as quickly as possible from the culture to the respirometer. The respiration rate of the carbon-limited bacterial cells is greater than that of nitrogen-limited cells (Tempest *et al.*, 1967). It follows from the results of Tempest & Herbert (1965) and Harrison & Loveless (1971a) that the q_{O_2} of resting cells increases with the growth rate of the cultures from which the cells were taken. This effect suggests that the content of catabolic and respiratory enzymes in the biomass is continuously regulated by growth rate.

The q_{O_2} of resting bacteria may be markedly affected by the dissolved oxygen tension during growth. With *Escherichia coli* the maximum q_{O_2} was obtained with oxygen-limited growth of the cells but, paradoxically, with *Klebsiella aerogenes* the maximum q_{O_2} was obtained when the organisms were grown under anaerobic conditions (Harrison & Loveless, 1971a).

11.5 Influence of dissolved oxygen tension on amounts of respiratory and catabolic enzymes in cells

In eucaryotes and in some bacteria the cytochromes form the terminal oxidative pathway leading to oxygen. In some bacteria, for example among the lactobacillaceae, there is a more obscure pathway to oxygen via flavoproteins. Harrison (1972) found that the terminal oxidase in *Klebsiella aerogenes*, cytochrome a_2, increased 200-fold in amount on lowering the DOT from 5·3 mmHg to <0·4 mmHg; the amounts of cytochromes *b* and *o* were little affected by the DOT value. In a *Candida* culture cytochromes *b* and *c* showed maxima at a DOT value about 1 mmHg and cytochrome *a* showed a maximum at a DOT <0·1 mmHg (Moss *et al.*, 1969).

Carter & Bull (1969) found that the amounts of some enzymes in the glycolytic and hexose monophosphate pathways reached maxima at DOT values below 30 mmHg in cultures of *Aspergillus nidulans*.

11.6 Transitions between aerobic and anaerobic metabolism in facultative anaerobes

On transition from anaerobic to aerobic growth synthesis of many different enzymes is induced and synthesis of enzymes required for anaerobic growth is repressed (Wimpenny, 1969). Among the enzymes induced by aerobiosis are components in the cytochrome pathway, including ubiquinone, and citric acid cycle enzymes, whilst synthesis of formic hydrogenlyase is repressed.

In the oxygen-limited state the metabolism of a facultative anaerobe such as *K. aerogenes* can be a combination of aerobic and anaerobic metabolism. At an acid pH these bacteria under oxygen limitation produce exclusively 2,3-butanediol and carbon dioxide from glucose; excess oxygen suppresses butanediol production, and the anaerobic condition causes accumulation of ethanol as well as butanediol (Pirt & Callow, 1958a).

Harrison & Pirt (1967) tried to determine the DOT value at which transition from aerobic to anaerobic metabolism occurs in cultures of *K. aerogenes* growing on glucose. They found that down to a DOT value of 5 to 15 mmHg the q_{O_2} was independent of DOT but further lowering of the DOT caused a sudden stimulation of respiration rate increasing it by about 30%; this led to oscillations in the DOT and in q_{O_2}. Thus instability in regulation of q_{O_2} was the first sign of oxygen limitation in the growing culture.

The results of Harrison & Loveless (1971b) showed that the time required for the transition from steady-state anaerobic growth to the steady state of aerobic growth in cultures of *K. aerogenes* or *E. coli* was about twice the doubling time; for the reverse change, from aerobic to anaerobic state, about three times the doubling time was required.

11.7 Effects of dissolved oxygen tension on functions other than respiration

The viability of acetic acid bacteria has been found to require the presence of oxygen; withdrawal of oxygen from a growing culture for as little as one minute caused extensive death of the bacteria (Hromatka, 1952).

Feren & Squires (1969) found that to maintain maximum production of the antibiotic cephalosporin c, the mould *Cephalosporium* required a DOT not less than 20 mmHg whereas the critical DOT for respiration was 8 mmHg. There has been little systematic investigation of the effect of dissolved oxygen concentration on production of antibiotics and other secondary metabolites. Definitive knowledge of these effects can probably only be obtained by means of steady-state chemostat cultures. Some examples of the application of this technique are given below.

The production of the insoluble pigment melanin in the mycelium of *Aspergillus nidulans* was found to be maximal when the DOT was > 30 mmHg. At

DOT values < 30 mmHg a soluble form of the pigment was secreted into the medium and reached a maximum amount at DOT 18 mmHg (Rowley & Pirt, 1972). Production of the diphtheria toxin by corynebacteria is maximal at DOT values from < 0·1 mm to 100 mmHg; higher DOT values inhibited the toxin production (Righelato & Van Hemert, 1969). The composition of the lipids in *Candida utilis* is dependent on the DOT during growth. Decreasing the DOT from 5 to < 1 mmHg decreased the ratio of $C_{18}:C_{16}$ fatty acids and decreased the degree of unsaturation of the fatty acids (Brown & Rose, 1969). The decrease in the unsaturation may be attributed to replacement of linolenic acid, which has 3 double bonds, by oleic acid which has one double bond (Babij *et al.*, 1969). Harrison *et al.* (1969) reported that the production of a yellow flavin-like pigment by *Klebsiella aerogenes* increased with DOT value up to at least 450 mmHg. This result indicates that it would be of interest to study the effects of hyperbaric oxygen tensions on product formation.

11.8 Substitutes for oxygen

Oxygen may be supplied to a culture in the form of hydrogen peroxide, a process which results in the induction of an extremely high content of catalase in the organism (Herbert & Phipps, 1974). If oxygen is absent, nitrate, ferricyanide, tetrathionate and some organic reducible compounds such as methylene blue and tetrazolium may act as final electron acceptors. Oxygen inhibits the cell's reaction with alternative electron acceptors. Nitrate reduction like that of oxygen is coupled to ATP production in *Klebsiella*, with glucose as energy source. The stoichiometry is $NO_3^- \rightarrow NO_2^- + 3$ ATP (Hadjipetrou & Stouthamer, 1965). The process differs from that with oxygen in that the citric acid cycle enzymes are repressed so that glucose is oxidized only as far as acetate.

With ferricyanide in place of oxygen in *Klebsiella* cultures, the citric acid cycle enzymes are not repressed, so that glucose is completely oxidized; however, the ferricyanide reduction is not coupled to ATP production (Hadjipetrou *et al.*, 1966). Thus growth of *Klebsiella* on glucose in the presence of ferricyanide depends on the ATP produced in glycolysis. Both nitrate and ferricyanide, like oxygen, repress synthesis of the enzyme formic hydrogenlyase.

11.9 Inhibition by oxygen

There are few measurements of the inhibitory levels of dissolved oxygen; however, these indicate that for all organisms there is a level of DOT above which oxygen becomes toxic and inhibits growth. The different types of response are shown in Fig. 11.3. For anaerobes, oxygen at all values of DOT is inhibitory

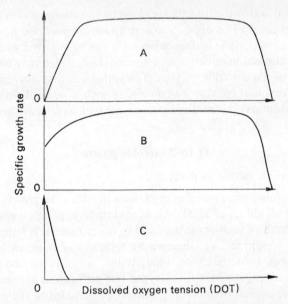

FIG. 11.3   Responses of microbial growth rate to dissolved oxygen tension (DOT):
  A, aerobic organism, B, facultative anaerobe; C, anaerobic organism.

to growth (Fig. 11.3c). Facultative anaerobes are distinguished by being able
to grow either with or without oxygen. The term *microaerophilic* may be
applied to aerobes or facultative anaerobes for which the inhibitory DOT is
less than 0·21 atm, that is the oxygen partial pressure in air at 1 atm. For other
aerobic organisms the inhibitory DOT is hyperbaric, that is >0·21 atm. Most
studies of the effects of hyperbaric oxygen have been made with control of the
oxygen tension in the gas phase rather than DOT in the culture medium.
Such studies are inconclusive about the quantitative relation between DOT
and its effects on growth, unless precautions are taken to ensure that the decrease in DOT due to oxygen uptake is insignificant. When this was done with
Escherichia coli colonies growing under glucose-limited conditions on agar it
was found that the growth of *E. coli* is prevented by a DOT of 1 atm (Pirt,
1967). The results of Wiseman *et al.* (1966) indicate that growth of *Pseudomonas*, *Escherichia* and *Staphylococcus* is prevented by an oxygen tension of
about 1 atm. Growth of *Penicillium chrysogenum* is inhibited by a DOT of
1·5 atm and, in a stirred fermenter, such a high oxygen tension caused the
hyphal morphology to be aberrant (Pirt, unpublished work). Thus the inhibitory DOT for aerobes may be sufficiently low for oxygen toxicity to develop in deep industrial scale fermentation tanks especially if positive pressure
is maintained in the head space.

The molecular basis of oxygen toxicity is obscure. Various hypotheses have
been proposed to account for the effects but there is no unifying hypothesis.

One hypothesis is that the toxicity is caused by the accumulation of hydrogen peroxide by the cells. This appears to happen in *Streptococcus faecalis* (Seeley & Vandemark, 1951). The hydrogen peroxide can be removed and the toxicity overcome by the action of catalase or a peroxidase, which may be induced by the presence of the peroxide. A second hypothesis is that oxygen directly inactivates an essential enzyme by combining with it or oxidizing it. A third hypothesis is that oxygen oxidizes a coenzyme such as ferredoxin and thereby inactivates it.

11.10 Anaerobic growth

11.10.1 Anaerobic culture methods

Obligate anaerobes are defined as organisms in which the growth is inhibited by oxygen at all values of DOT. Air at atmospheric pressure completely inhibits the growth of such organisms and to many types it is lethal.

Anaerobic conditions are obtained by a variety of more or less exacting means (Hobson, 1969; Hungate, 1969; Willis, 1969). It may be sufficient to substitute the air atmosphere with oxygen-free nitrogen or carbon dioxide. In any case some carbon dioxide (5%) should be included in the gas phase. A hydrogen–carbon dioxide mixture is convenient in the 'anaerobic jar' in which the residual oxygen can be removed by catalytic combination with hydrogen. Anaerobic conditions are also achieved by adding to the medium reducing agents, particularly cysteine, sodium sulphide, sodium thioglycollate, sodium dithionite and ascorbic acid, the generally preferred agents being cysteine and sodium sulphide. Certain natural media such as cooked meats form suitable substrates possibly because they contain reducing substances. Prior growth of a facultative anaerobe in the medium has been recommended as a means of achieving anaerobiosis for the culture of methanogenic bacteria (Hobson, 1969).

The degree of anaerobiosis is often determined by E_h measurement. Redox dyes may be incorporated in the medium for this purpose. With resazurin (10^{-4} g/l), a commonly used E_h indicator, the E_0 for the change from pink to colourless is -51 mV at pH 7 and 30°C; the colourless state which occurs at an E_h of about -100 mV is taken to indicate anaerobic conditions.

11.10.2 Limits of E_h for anaerobic growth

The upper limit of E_h for growth of an anaerobe was first estimated by Knight & Fildes (1930) from observations on the relation between E_h and the germination period of spores of *Clostridium tetani*. The upper limit of E_h was 50 to 100 mV. Aubel *et al.* (1946) found that the upper limit of E_h for growth of *Cl. sporogenes* or *Cl. saccharobutyricum* was about -100 mV; however, the initial E_h could be up to $+180$ mV, but then there was a lag in growth until the organisms had spontaneously decreased the E_h to -100 mV. The initial

E_h limit tolerated was found to decrease with decrease in the inoculum size (expressed in opacity units by Aubel *et al.*, 1946). Also the supernatant fluid from a fully grown culture was found to decrease the minimum inoculum size which could grow with an initial E_h value of 180 mV. The results of Aubel *et al.* (1946) show that the clostridia possess a mechanism for decreasing the E_h value of the medium.

The lower limit of E_h reached in an unaerated bacterial culture is characteristic of the organism (Jacob, 1970). In cultures of *Bacillus subtilis* the E_h relative to the standard hydrogen electrode (Jacob gives values relative to the calomel electrode) fell from 400 mV at inoculation to 200 mV; in a culture of *Staphylococcus aureus* the E_h fell from 400 to 50 mV and in cultures of *Proteus vulgaris* and *Escherichia coli* the E_h fell from $+400$ to -300 mV. In contrast, cultures of *Clostridium paraputrificum* decreased the E_h value from about $+300$ mV to -300 mV, but this change took place during the lag before growth (Fig. 11.4). These results show that the facultative enterobacteria can decrease the E_h to a value which permits growth of a strict anaerobe.

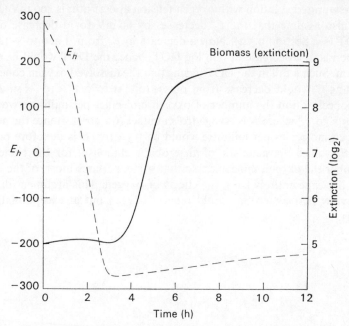

FIG. 11.4   Change in E_h and biomass growth (extinction) in culture of *Clostridium paraputrificum* in liver broth. (Redrawn from Jacob, 1970; permission of Academic Press).

11.10.3 Upper limits of dissolved oxygen concentration for anaerobic growth

Gordon *et al.* (1953) determined the tolerated oxygen partial pressure in the gas phase over solid media (blood agar) used for the culture of some clostridia.

The upper limit was found to be 30 to 80 mmHg for *Clostridium welchii*, 4 mmHg for *Cl. sporogenes* and 1 to 2 mmHg for *Cl. tetani*, *Cl. botulinum* and *Cl. oedematiens*. The oxygen tension of the medium may be lower than that in the gas phase because at least some clostridia have a mechanism for oxygen uptake. Bromel & Teodoro (1966) showed that *Cl. sporogenes* had a q_{O_2} of 30 ml oxygen/(g dry weight h) with glucose as substrate. This respiration of clostridia may be the cellular mechanism which lowers the DOT, and consequently the E_h, to a level which permits growth. The upper limit of E_h or DOT which permits growth may be a reflection of the organism's ability to remove oxygen from the medium and generate anaerobic conditions. Thus the most exacting anaerobes could be those with no mechanism for oxygen uptake.

Harrison & Pirt (1967) found that during anaerobic growth of *Klebsiella aerogenes* the dissolved oxygen tension was too low to measure, that is < 0·1 mmHg.

We may estimate the dissolved oxygen concentration in an 'anaerobic' culture from the E_h value in the following way. Suppose that the E_h of an oxygen-saturated solution with oxygen tension at 0·21 atm is 400 mV (Jacob, 1970); also assume that the E_h decreases by 60 mV for a decrease of 1 in log DOT (see Section 9.5.2). Now a decrease in E_h from +400 to −140 mV corresponds to a decrease of 9 in log DOT, hence the DOT would be 0·21 × 10^{-9} atm. Such a fall in the DOT means that the dissolved oxygen concentration at 25°C would decrease from 0·25 × 10^{-3} M to 0·25 × 10^{-12} M. At the latter concentration the number of oxygen molecules per millilitre would be N × 0·25 × 10^{-15} where N is Avogadro's number (6 × 10^{23}). Hence the number of oxygen molecules per millilitre would be 1·5 × 10^8. It is therefore possible at the E_h values characteristic of anaerobic metabolism for the bacteria to outnumber the oxygen molecules, so that some, perhaps most, of the organisms will escape entirely from the effects of oxygen. This argument disposes of the view that anaerobes could require oxygen but at exceptionally low tensions.

CHAPTER 12

General nutrition

12.1 Introduction

The nutrients required for growth, apart from energy sources, may be classified into the following groups: (*i*) sources of the 'major' elements C, H, O and N; (*ii*) sources of the 'minor' elements P, K, S, Mg; (*iii*) vitamins and hormones; (*iv*) sources of the 'trace' elements. The term *growth factor* is used to refer to essential organic nutrients such as amino acids, which are incorporated whole in the cell structure. This definition excludes carbon and energy sources because they are catabolized. Carbon and energy sources, which often are also sources of cell oxygen and hydrogen, are discussed in Chapter 8. In the present chapter, the nitrogen sources and the other groups of nutrients are discussed.

Microbes and tissue cells were first cultivated in 'natural' media which are defined as extracts of plant or animal material; examples are grape juice, milk, corn steep liquor, peptone and serum. Such media are often convenient because they are sources of all four groups of nutrients, but they suffer from the disadvantages of being undefined and variable in composition. In order to elucidate nutritional effects, as far as possible, chemically defined or 'synthetic' media must be used. The first steps towards the definition of culture media were made by Pasteur (1869) who introduced a partially defined medium—which consisted of glucose, ammonium tartrate and ash of yeast—for the growth of yeast. The role of the tartrate is not clear but it may have made the medium selective for yeast growth since at that time there was no other means of obtaining a pure culture. A necessity for vitamins in microbial nutrition was first demonstrated for yeast growth by Wildiers (1901), who showed that the defined medium of Pasteur was inadequate for the growth of yeast unless an organic extract of yeast cells, termed *bios*, was included in the medium. Subsequently it was found that the bios factors are water-soluble vitamins. How did Pasteur succeed in growing yeast in his defined medium without apparently adding the bios factors? One can surmise that it was the result of using a heavy inoculum, which contained an excess of the growth factors, or through working with a mixed culture of species which between them could synthesize the growth factors.

117

The first fully defined medium for microbial growth was that developed for the culture of *Aspergillus niger* by Raulin (1869), who was a student of Pasteur. The Raulin medium provided the major and minor elements and three trace elements (Fe, Zn and Si) all in inorganic form together with sugar as the source of C, H, O and energy. Subsequent work has confirmed Raulin's findings except for the requirement for Si in the form of silicate. Raulin defined not only the qualitative requirements but also the quantitative requirements for each nutrient in terms of the growth yield. It is unfortunate that most subsequent workers have neglected to determine the growth yields from nutrients, which are essential to define quantitatively the nutrient requirements. For instance, neglect of this quantitative aspect has accounted for much of the impasse in animal cell nutrition (Birch & Pirt, 1971).

A 'minimal' medium contains only the nutrients essential for growth. A 'rich' medium is one in which the essential nutrients are supplemented with other nutrients which act as alternative sources of the elements, usually in the form of amino acids, vitamins, nucleic acid precursors and other intermediates in cell synthesis. Enrichment of a culture media may increase the biomass growth rate and alter the enzyme composition of the biomass.

The nutrient requirements of an organism may vary qualitatively or quantitatively with the conditions of culture. Temperature can affect the growth factor requirements (Section **13.4**) and we can expect nutrient requirements to be dependent on pH value and on the osmolality of the medium.

Initiation of growth from small inocula sometimes requires a richer medium than does growth with high population densities. For example, *Klebsiella aerogenes* fails to grow in minimal glucose-ammonia media when the inoculum size is less than about 10^5 bacteria/ml, however, addition of an amino acid (asparagine or alanine) lowers the minimum inoculum density required (Lodge & Hinshelwood, 1943).

12.2 Microbiological assay of growth factors

The microbiological assay of amino acids, vitamins and other substances which can act as growth factors depends upon the constancy of growth yields or product yields, which are defined in Chapter 2. The assay material is made the growth-limiting substrate with all other required substrates in excess. Ideally a straight-line relation should be obtained between the amounts of the biomass formed and the substrate assayed as shown in Fig. 12.1. If x_1 = biomass in v ml of standard solution of growth factor of concentration c_1, x_2 = biomass in v ml of assay sample of concentration c_2, Y = growth yield from assay material, then we have

$$x_1 = x_B + Yc_1v \qquad\qquad 12.1$$

$$x_2 = x_B + Yc_2v \qquad\qquad 12.2$$

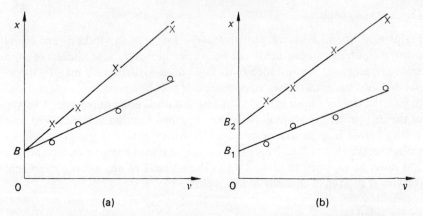

FIG. 12.1 Microbiological assay of a growth factor by measurement of maximum
biomass produced (x); $v =$ volume of sample. (a) Valid assay; (b) non-valid assay.
Circles, standard samples; crosses, test samples; B, B_1, B_2 are intercepts.

where x_B is the blank value due to the inoculum (x_0) plus any assay material
(s_B) present as an impurity in the assay medium, that is

$$x_B = x_0 + Ys_B \qquad\qquad 12.3$$

The ratio of the slopes of the graphs of Eqn 12.1 and 12.2 will be c_1/c_2, hence
c_2 can be calculated.

For a valid assay the growth yield must be the same with standard and test
solutions. If this is so, the value of the blank growth Ys_B in Eqn 12.3 will be
the same for test and standard assays, that is the intercepts on the ordinate
will be equal as shown in Fig. 12.1(a). If the yields with different test samples
are not equal, possibly because there is present in the test sample some additional material which has a sparing action on the growth factor, then the assay
will not be valid. Invalidity of the assay is shown by inequality of the intercepts (Fig. 12.1b), in which case the intercepts are $x_{B1} = x_0 + Y_1 s_B$ and
$x_{B2} = x_0 + Y_2 s_B$ where Y_1 and Y_2 are different growth yields. This test of
validity requires that $s_B > 0$.

12.3 Nitrogen requirements

12.3.1 General

The nitrogen sources which can be utilized by different organisms probably
include most, if not all, of the inorganic and organic forms of nitrogen. The
nitrogen is metabolized to provide mostly protein, nucleic acids and cell wall
polymers. The amino acid pool in the cytoplasm accounts for 0·25 to 5% of
the dry biomass (Mandelstam, 1958; Brown & Stanley, 1972). Much of the
pool consists of glutamate. The cell nitrogen constitutes up to 12% of the dry
weight of bacteria and 10% of fungal dry weight.

12.3.2 Amino acids

Frequently, amino acids are growth factors. For protein synthesis and many other purposes L-amino acids are required; however, sometimes D-amino acids are required, for instance D-alanine and D-aspartic acid may be incorporated into bacterial cell walls. Amino acids may be racemized in the biomass to give the isomer required. Individual amino acids constitute about 1 to 5% of the cell protein, on which basis the amount of amino acid required as a growth factor may be estimated roughly. An exception is glutamic acid or glutamine which quantitatively plays the major role in amino acid metabolism and must be supplied in several times the amount of any other amino acid (Davies *et al.*, 1965; Griffiths & Pirt, 1967).

12.3.3 Peptides

Some bacteria, which require a number of amino acids, grow faster when provided with one or more of the amino acids in peptide form, for example histidine peptides for *Lactobacillus delbrueckii* (Peters *et al.*, 1953). *Lactobacillus casei* growing in the absence of pyridoxin requires both D- and L-alanine, however, D-alanine antagonizes the uptake of L-alanine and the antagonism can be overcome by supplying L-alanine in peptide form (Kihara & Snell, 1952). Also in *L. delbrueckii* the uptake of serine is antagonized by both D- and L-alanine and the antagonism is overcome by supplying serine peptides (Prescott *et al.*, 1953). In *Streptococcus faecalis*, tyrosine is decarboxylated by the cells in culture to such an extent that tyrosine deficiency can occur. Tyrosine dipeptides effectively provide tyrosine for these bacteria (Kihara *et al.*, 1952).

Frequently essential amino acids act as growth inhibitors. Inhibition by the amino acids is classed as the competitive type if it is antagonized by some other amino acid, or as the non-competitive type if it cannot be overcome by other amino acids. Antagonisms have been observed between uptake of amino acids in the following groups: (*i*) phenylalanine, tyrosine, tryptophan; (*ii*) serine, threonine, alanine, glycine; (*iii*) glutamic acid, aspartic acid (Snell, 1949; Lichstein, 1960); (*iv*) valine, leucine, isoleucine (Kepes & Cohen, 1962, p. 210); (*v*) norleucine, methionine (Kepes & Cohen, 1962, p. 211). These antagonisms are attributed to competition for a common permease. In *Cephalosporium*, nitrate ions antagonize the uptake of methionine by an auxotrophic strain and ammonium ions overcome the antagonism (Nüesch, private communication). Histidine inhibition of the growth of a *Bacillus subtilis* mutant was overcome by a glycine tripeptide (Demain & Hendlin, 1958). For the same organism enzyme-hydrolysed casein became inhibitory to growth at concentrations above about 100 μg/ml. Non-competitive inhibition of growth by amino acids as well as other intermediates in metabolism is commonly found in autotrophic microorganisms (Kelly, 1967, p. 47) and in bacterial growth on other C_1 compounds such as methane (Eroshin *et al.*, 1968).

12.3.4 Nitrogen-limited growth

The protein content of the biomass is lower with nitrogen-limited growth than with carbon-limited growth, for example in yeast there is 30% protein with ammonia limitation compared with 50% with glycerol limitation (Light, 1972). The low protein content with excess carbon source could reflect the accumulation of energy reserves such as glycogen in the biomass.

12.4 Vitamin and hormone requirements

The term *vitamin* is used here to refer to growth factors other than amino acids. The vitamins are classed into two groups: the fat-soluble, and the water-soluble. The fat-soluble group consists of vitamins A, D, E, K, ubiquinone, cholesterol and the unsaturated fatty acids (oleic, linoleic, linolenic and arachidonic acids). Vitamins A, D and E, which are important in human or animal nutrition, have not been found necessary for growth of any protists. Unsaturated fatty acids are required by some lactobacilli (Snell, 1949) and by a *Sarcina* species (Lichstein, 1960). The fatty acids often are conveniently supplied in the form of water-soluble esters (the Tween compounds). Cholesterol is required by many of the mycoplasmas.

The water-soluble vitamins are: ascorbic acid, thiamine, riboflavin, pantothenic acid, pyridoxin, nicotinic acid, biotin, *p*-aminobenzoic acid, folic acid, cobalamin, mevalonic acid, choline and meso-inositol. Various derivatives or 'vitamers' exist for some of the vitamins, for example pyridoxamine and pyridoxal are vitamers of pyridoxin, and the organism may require one specific vitamer. All of the water-soluble vitamins, with the exception of ascorbic acid, have been found to be growth factors for some protists and animal tissue cells in culture. Most of the water-soluble vitamins are components of coenzymes. Choline and inositol are constituents of lipids. Of the inositol isomers only meso- (or myo-) inositol has growth factor activity. Provision of the end products of metabolic pathways in which the vitamin is involved may substitute for, or spare, the vitamin. A case in point is substitution of unsaturated fatty acids for biotin (Koser, 1968). In undefined media, yeast extract is commonly used as a source of vitamins.

Vitamin requirements for microbial cultures have rarely been defined quantitatively in terms of the growth yield so that the amount required to produce a given amount of biomass is unknown. Some data on growth yields from vitamins when they are supplied in excess are given in Table 12.1. The growth yield from a vitamin is often higher when it is growth-limiting than

TABLE 12.1 Growth yields from vitamins

Vitamin	Protist or cell type	Growth yield* (g dry biomass/g vitamin)
Biotin	Streptococcus[1]	$1 \cdot 08 \times 10^6$
	Mouse LS cell[2]	$2 \cdot 0 \times 10^5$
Folic acid	Mouse LS cell[2]	$2 \cdot 4 \times 10^5$
Riboflavin	Streptococcus[1]	$7 \cdot 4 \times 10^4$
	Mouse LS cell[2]	$2 \cdot 0 \times 10^4$
Pantothenic acid	Streptococcus[1]	$1 \cdot 53 \times 10^4$
	Mouse LS cell[2]	$6 \cdot 0 \times 10^3$
Thiamine	Streptococcus[1]	$1 \cdot 16 \times 10^4$
	Mouse LS cell[2]	$3 \cdot 2 \times 10^4$
Nicotinic acid	Streptococcus[1]	$3 \cdot 4 \times 10^3$
Nicotinamide	Mouse LS cell[2]	$4 \cdot 8 \times 10^2$
Pyridoxin	Mouse LS cell[2]	$4 \cdot 36 \times 10^2$
Mesoinositol	Mouse LS cell[2]	$1 \cdot 56 \times 10^2$
Choline	Mouse LS cell[2]	$0 \cdot 71 \times 10^2$

[1]Carlson (1971), [2]Blaker & Pirt (1971), Blaker (1971)

* Growth yields in presence of excess vitamin

when it is supplied in excess. The growth yield may vary inversely as the growth rate, for instance the growth yield of a yeast from thiamine decreased from $29 \cdot 0 \times 10^5$ to $1 \cdot 2 \times 10^5$ g dry biomass/g thiamine when the specific growth rate was increased from $0 \cdot 3$ to $0 \cdot 8$ times the maximum value (Button, 1969). The biotin requirement of *Corynebacterium glutamicus* is decreased by 90% when the bacteria are grown on acetate instead of glucose (Kinoshita, 1972, p. 271), hence the nature of the carbon and energy source can affect the vitamin growth yield.

The effects of making vitamins growth-limiting have not been studied systematically but there is evidence that the effects can be dramatic, for instance a biotin deficiency is one of the factors responsible for the over-production of glutamic acid by *Corynebacterium glutamicus*. The deficiency apparently causes an increase in the membrane permeability to glutamic acid so that the amino acid cannot be retained in the cell (Kinoshita, 1972, p. 314). In cultures of mouse LS cells growth limitation by choline or inositol results in the formation of budlike protuberances on the cells, which suggests that the membranes are defective (Blaker, Tovey & Pirt, unpublished).

There can be antagonisms between vitamins, for instance between thiamine and pyridoxal in a yeast (Snell, 1949). Little is known about growth inhibition by vitamins but folic acid or folinic acid at $0 \cdot 01$ μg/ml completely prevented growth of *Lactobacillus bulgaricus* (Rogosa et al., 1961).

Any cell constituent may be anticipated to occur as a growth factor. Other growth factors of importance are purines, pyrimidines, haemin (Koser, 1968) and the polyamines, putrescine, spermine and spermidine (Cohen, 1971).

Studies on the role of the hormone insulin required in human HeLa cell culture (Blaker *et al.*, 1971) and on the hormone 2,4-dichlorophenoxyacetic acid required in plant cell culture (Yasuda *et al.*, 1972) show that these hormones act like growth factors in the culture media. The insulin requirement of HeLa cells could easily be missed since when the cells are taken from complete medium they do not show the sign of insulin deficiency (a marked fall in the growth rate) until after the thirteenth generation. Also this result suggests that a growth factor deficiency may sometimes be indicated by a decrease in growth rate rather than a decrease in growth yield.

12.5 Phosphorus requirements

Phosphorus is usually supplied in the form of inorganic phosphate. Alternatives are organic phosphates such as glycerophosphate and phospholipid. The phosphate is mostly incorporated into the nucleic acids, phospholipids and cell wall polymers; occasionally it may be stored in the cell as polymetaphosphate (Markham & Byrne, 1969). Only a small fraction of the total phosphate appears in the form of diffusible organic phosphates such as ATP.

The phosphorus content of bacterial cells is about $1 \cdot 5\%$ of the dry biomass, however, the content increases with growth rate and varies inversely with temperature. This variation largely reflects the RNA content of the cells (Tempest, 1969). In bacteria there is a stoichiometry between the amounts of magnesium, potassium, phosphate and RNA which is characteristic of different groups of bacteria (Tempest, 1969). In Gram-negative bacteria the molecular ratio of $Mg:K:RNA$ nucleotide: PO_4 is close to $1:4:5:8$ and independent of the growth rate, temperature and growth-limiting substrate. In contrast, in Gram-positive bacteria, the ratio of $Mg:K:RNA$ nucleotide: PO_4 is about $1:13:5:13$, except when phosphate limits the growth when the ratio is about $1:4:5:8$, i.e. the same as for Gram-negative bacteria. The higher phosphate and potassium contents of the Gram-positive bacteria are accounted for by the presence of teichoic acid in the cell walls when phosphate is in excess. However, with phosphate-limited growth, the teichoic acid polymer is replaced by the phosphate-free teichuronic acid.

12.6 Potassium and sodium requirements

The potassium requirement for the growth of microbial biomass corresponds to a growth yield of roughly 60 g dry biomass/g potassium. Much of the potassium seems to be bound up with the RNA (Tempest, 1969) so that the potassium requirement is increased by factors such as growth rate which increase the RNA content of the biomass. The potassium requirement may vary inversely as the pH value since, in *Klebsiella*, the potassium content increased about 30% on decreasing the medium pH from 7 to 6 (Eddy *et al.*, 1951). The potassium content of some bacteria is increased up to threefold by increasing the sodium chloride content of the medium (Tempest & Meers, 1968). For growth of mammalian cells in culture a threshold value of potassium ion concentration (4×10^{-4} M) had to be exceeded before growth could occur and for maximum growth rate the potassium ion concentration had to be at least 5.3×10^{-4} M (Birch & Pirt, 1971). Threshold concentrations for nutrients seem, so far, to be unique to potassium and magnesium ions in mammalian cell growth. The only other alkali metal which can replace potassium is rubidium (Eddy & Hinshelwood, 1951) though this reduces the maximum growth rate. The effect of potassium ions may be antagonized by ammonium ions (Dicks & Tempest, 1967). Potassium ions function as coenzymes and probably act as cations in the structure of RNA and other anionic structures in the cell.

A requirement for sodium ions in microbial growth has rarely been demonstrated, but this may be largely because it is difficult to produce media free from sodium ions. Large amounts of sodium ions may be required by halophilic bacteria. The requirement for sodium as a trace element is referred to in Section 12.9.2.

12.7 Magnesium requirements

The growth yield from magnesium ions varies from about 300 to 900 g dry biomass/g magnesium and is inversely proportional to the amount of RNA in the biomass (Tempest, 1969). In algae, magnesium is present in the chlorophyll. Birch & Pirt (1971) observed that for mammalian cells in culture there is for magnesium ions, as for potassium, a threshold concentration (5×10^{-5} M Mg^{2+}) which must be exceeded for growth to occur, and for maximum growth rate a concentration of 2×10^{-4} M Mg^{2+} is necessary.

A unique feature of magnesium deficiency in a chemostat culture of *Agrobacterium tumefaciens* is that it causes large oscillations in the biomass concentration reflecting oscillations in the growth rate (Kurowski *et al.*, 1973).

12.8 Sulphur requirements

The growth yield from assimilated sulphur is about 300 g dry biomass/g sulphur. Sulphur is commonly provided in one of its inorganic forms, usually sulphate, or in the form of cysteine or methionine. Under aerobic conditions cysteine is almost entirely converted to cystine, however, the latter compound usually can substitute for cysteine.

Sulphur sources are assimilated mostly to provide sulphur for the amino acids L-cysteine and L-methionine. Smaller amounts of sulphur sources are required to provide the sulphur groups in some coenzymes such as biotin, coenzyme A, ferredoxin, lipoic acid and thiamine. Growth limitation by sulphur may decrease the synthesis of these key sulphur compounds and affect their functions in the cell. An example is sulphate-limited growth of yeast, which causes loss of one site of oxidative phosphorylation in the respiratory chain, the effect being similar to that of iron-limited growth (Light, 1972).

12.9 Trace elements

12.9.1 Effects in general

The trace elements found essential for growth or for which a requirement has been suggested are given in Table 12.2. These elements fall between atomic numbers 4 (beryllium) and 74 (tungsten) but nearly all are in the series from atomic number 4 to 35. In general, the requirements for trace elements are only known qualitatively and for this reason they are almost invariably added to media in arbitrary amounts. It is usually difficult to demonstrate a requirement for the trace elements because often they are present in sufficient amount as contaminants in the medium constituents. Attempts to determine the growth yields for trace elements suggest that these vary widely with the growth rate and other conditions, for example see Light (1972). Rough estimates of the quantitative requirements for some of the more important trace elements are (in g element/100 g dry biomass): Ca 0·10, Fe 0·015, Mn 0·005, Zn 0·005, Cu 0·001, Co 0·001, Mo 0·001.

In batch cultures trace element deficiency probably is reflected more by limitation of growth rate than by limitation of biomass concentration as shown in Fig. 12.2(a). Such an effect was found with iron-deficient growth of *Agrobacterium tumefaciens* (Kurowski & Pirt, 1971) and iron-deficient growth of mammalian cells (Birch & Pirt, 1970). In a chemostat culture a trace element deficiency is shown by decrease in the steady-state biomass with increase in growth rate, that is the growth yield varies inversely with the growth rate without alteration in the critical dilution rate (Fig. 12.2b). Manganese deficiency in a culture of *Agrobacterium* affects growth in the ways indicated in Fig. 12.2(a) and (b) (Kurowski & Pirt, unpublished). The requirement for a trace element, it has been suggested (Hutner, 1972, p. 338) may increase several fold when the culture is subjected to stress such as an increase in temperature above the optimum value. This may provide a technique for determining more easily the qualitative trace element requirements of a culture. Very little is known about the toxic concentrations of trace elements although they are generally considered to be of the order 10^{-4} M. Metal ion antagonisms may occur, for example between Mn^{2+} and Zn^{2+} in *Rhizobium*, species (Wilson & Reisenauer, 1970). Also the effects may not be 'mono-specific',

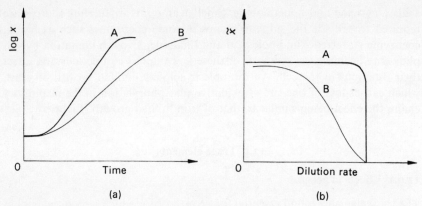

FIG. 12.2 Effects of trace element deficiency on growth of microorganisms in
(a) batch culture and (b) chemostat culture: x = biomass in batch; $\tilde{x}$ = steady state
biomass in chemostat. Line A, optimum amount of trace element; line B, deficiency
of trace element.

thus several different ions may qualitatively act in the same way (Dixon &
Webb, 1967, p. 421).

12.9.2 Specific effects of trace elements

The group A trace elements given in Table 12.2 are frequently found essential
for growth. Calcium is the cation in the dipicolinate salt present in spores of
Bacillus species. Also calcium is a cofactor for the enzyme α-amylase.

TABLE 12.2 Trace elements which may be required in microbe and cell culture
(Hutner, 1972; Tempest, 1969; Steinberg, 1956; Arnon, 1938)

A	Elements which are frequently essential for growth
	Ca, Mn, Fe, Co, Cu, Zn
B	Elements which are, rarely, essential for growth
	B, Na, Al, Si, Cl, V, Cr, Ni, As, Se, Mo, Sn, I
C	Elements which may be, rarely, essential for growth
	Be, F, Sc, Ti, Ga, Ge, Br, Zr, W

Note: each group of elements is arranged in order of increasing atomic number

Manganese is generally considered to stimulate the growth of the lactic acid
bacteria and is involved in the sporulation of *Bacillus* species. In *Agrobacterium tumefaciens* a deficiency of manganese causes the transformation of
sucrose to 3-ketosucrose instead of its complete oxidation (Kurowski *et al.*,
1973). Iron and manganese deficiencies appear to cause accumulation of citric
acid by certain *Aspergillus niger* species (Choudhary & Pirt, 1966).

Iron is required for the many cofactors of enzymes with redox functions in the cell, for example the haem pigments of the cytochromes and catalase, ferredoxin and flavoproteins. The restriction of iron supply can disrupt the synthesis of these cofactors and impair their functions in the cell (Light, 1972). The iron-free cofactor may be excreted (Townsley & Neilands, 1957). Iron deficiency causes excretion of iron-binding compounds by some bacteria and fungi (Hutner, 1970; Snow, 1970). These iron-binding compounds chelate Fe^{3+} with high specificity (log stability constant > 30) and probably function as ionophores transporting iron across the plasma membrane of the cell (Ratledge, 1971). In *Corynebacterium diphtheriae*, iron deficiency is a necessary condition for excretion of the toxin (Righelato & Van Hemert, 1969a). Iron deficiency stimulates riboflavin production by certain yeasts (Demain, 1972a), and it has been suggested (Demain, 1972b) that iron deficiency may cause a switch from iron-protein to flavoprotein pathways of electron transfer.

Cobalt forms a part of the molecule of vitamin B_{12} or cobalamin, which is synthesized by procaryotes.

Copper is present in the terminal oxidase of the respiratory pathway in yeast. A deficiency of copper or iron causes enlargement of mitochondria in yeast (Davison *et al.*, 1972); also copper deficiency in a yeast culture selects a mutant which lacks the normal terminal oxidase (Downie & Garland, 1972) and uses a modified respiratory pathway. Peptidases may contain either copper or zinc ions as cofactors.

The group B trace elements (Table 12.2) are those for which usually no requirement can be shown but occasionally are required in specialized functions. Borate was found essential for growth of a *Candida* species on *n*-paraffins (Sato *et al.*, 1972). *Rhodopseudomonas spheroides* required Na^+ in trace amounts for its growth (Sistrom, 1960). O'Brien & Stern (1969) found that for the anaerobic growth of *Klebsiella aerogenes* on citrate there was a specific requirement for a massive amount of Na^+ (0·1 M). The sodium ion was required by oxalacetate decarboxylase, a requirement which disappeared during growth on glucose.

Selenite or selenide (but not selenate) is required for formation of formate dehydrogenase in *Escherichia coli* grown anaerobically, however, selenium deficiency had no effect on growth (Enoch & Lester, 1972).

Nickel ions are required by *Hydrogenomonas* oxidizing hydrogen for energy (Bartha & Ordal, 1965).

Molybdenum, supplied as molybdate, is a cofactor in bacterial nitrogenase which fixes nitrogen, and in nitrate reductase which is required for the assimilation of nitrate nitrogen by bacteria and fungi. Vanadium appears to be a less effective substitute for molybdenum.

Iodide ions are necessary for growth of a *Candida* species on *n*-paraffins (Sato *et al.*, 1972).

Knowledge of the requirements for trace elements and the effects of deficiencies will require improved means for the control of the trace element content of media.

12.10 Removal of trace elements from media

The trace metals may be removed from solution through precipitation as hydroxides, phosphates, or carbonates (Steinberg, 1956), and as ferrocyanides (Choudhary & Pirt, 1966). Other methods are solvent extraction with an organic complexant in the non-polar phase (Donald, 1952) adsorption on alumina (Ratledge & Chaudhry, 1971) and by chelating resins (Noguchi & Johnson, 1961). The effects of trace metal deficiency sometimes may be produced simply by adding a metal chelating agent to the medium, for example iron deficiency, which stimulates overproduction of riboflavin in some yeasts, is achieved by adding o-phenanthroline to the medium (Demain, 1972). Systematic methods for removing non-metals remain to be evolved.

12.11 Metal ion chelation

12.11.1 Introduction

Probably in all media, the concentrations of metal ions, other than alkali metal ions, are modified by chelation because many common medium constituents and culture products such as amino acids and hydroxy acids act as metal ion complexants. In order to prevent the precipitation of trace metal ions and to control their concentrations it is generally essential to chelate the metal ions by an added agent. Metal chelating agents, which are polybasic acids, such as ethylenediamine tetra-acetic acid, act as metal ion buffers.

12.11.2 Stability of metal chelates

The metal ion (M) combines reversibly with the ligand (L) according to the equation

$$M + L \rightleftharpoons ML$$

where the charges on the cation and ligand are omitted. The 'stability' constant of the complex is defined by

$$K = [ML]/[M][L] \qquad 12.4$$

where the square brackets denote concentration of the reactant. The stability constant is the reciprocal of a dissociation constant. The greater the value of

K the greater is the affinity of the ligand for the metal ion. Taking logarithms and putting pM $= -\log$ [M], Eqn *12.4* becomes

$$pM = \log K + \log \frac{[L]}{[ML]} \qquad\qquad 12.5$$

12.11.3 Effect of hydrogen ions on pM

A chelating agent can be represented as an acid H_mL where m is the number of hydrogen ions which can combine with the ligand, L. In effect, hydrogen ions compete with the metal ions for the ligand.

Let $L_u =$ sum of all forms of the complexant not combined with the metal ion, that is

$$L_u = L + HL + H_2L + \ldots + H_mL \qquad\qquad 12.6$$

where the charge on the ligand is omitted. It follows that

$$L_u = \alpha L \qquad\qquad 12.7$$

where $\alpha > 1$. The value of α can be expressed in terms of the stability constants of the acids HM, H_2M etc. and the concentration of hydrogen ions (Flaschka, 1964, p. 27). Substituting for [L] in Eqn *12.5* we obtain

$$pM = \log K - \log \alpha + \log \frac{[L_u]}{[ML]} \qquad\qquad 12.8$$

Putting $\log (K/\alpha) = \log K_{app}$, Eqn *12.8* becomes

$$pM = \log K_{app} + \log \frac{[L_u]}{[ML]} \qquad\qquad 12.9$$

In a strongly alkaline solution the value of $\alpha \approx 1$ but it increases with increasing acidity and at lower pH values it will substantially decrease the value of pM. The effect of pH on $\log \alpha$ for EDTA is shown in Fig. 12.3.

12.11.4 Calculation of pM

Let $[L_T] =$ total amount of complexant present (bound and unbound) and similarly let $[M_T] =$ total amount of metal ion present. When the metal ion is present in excess, that is $[M_T] > [L_T]$ for a powerful complexant we can put for the concentration of free metal ions

$$[M] = [M_T] - [L_T] \qquad\qquad 12.10$$

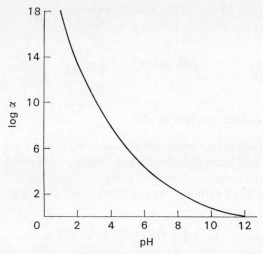

FIG. 12.3 The effect of pH value on the stability constants of EDTA complexes: $\log K_{app} = \log K - \log \alpha$. (Redrawn from Flaschka, 1964; permission of Pergamon Press)

When the complexant is in excess, that is $[L_T] > [M_T]$ we substitute in Eqn *12.9* $[L] = [L_T] - [M_T]$ and $[ML] = [M_T]$ and obtain

$$pM = \log K_{app} + \log \left\{ \frac{[L_T] - [M_T]}{[M_T]} \right\} \qquad 12.11$$

When the total amounts of complexant and metal ion are the same, that is $[L_T] = [M_T]$, since

$$[L_u] = [L_T] - \{[M_T] - [M]\} \qquad 12.12$$

it follows that

$$[L_u] = [M] \qquad 12.13$$

Substituting for $[L_u]$ in Eqn *12.9* and putting $[ML] = [M_T]$ we obtain

$$pM = \tfrac{1}{2}\{\log K_{app} - \log [M_T]\} \qquad 12.14$$

The change in pM with increasing amount of complexant and the effects of different stability constants are depicted in Fig. 12.4. With excess complexant the upper value of pM approximates to $\log K_{app}$. In the case of those ligands which form complexes of the form ML_2, the stability constant is given by β_2 (see Table 12.3) and the pM approaches $\log \beta_2$ with excess of ligand.

12.11.5 Effect of a second metal ion

A second species of metal ion will compete with the first one for the complexant. If K_x and K_y are the stability constants of the metal ion species M_x

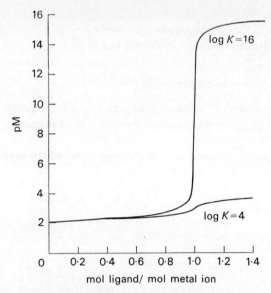

FIG. 12.4 Effect of increasing amount of chelating agent on the pM for two metal ions with different stability constants (K). Total metal ion present (free and bound), 10^{-2} M.

and M_y respectively then, equating concentrations of L from Eqn *12.4*, we have

$$\frac{[M_xL]}{[M_yL]} = \frac{K_x}{K_y} \cdot \frac{[M_x]}{[M_y]} \qquad 12.15$$

This equation shows that the two metal ions will compete effectively for complexant only if $K_x \approx K_y$. If $K_x \gg K_y$ and the total amounts of each metal ion and the ligand are about the same, virtually only M_x will be complexed. Also, as the amount of complexant in the mixture of metal ions is increased the two species of metal ions will be complexed in sequence, the one with the highest stability constant first.

12.11.6 Effect of a second ligand

Culture media often contain more than one constituent capable of chelating metal ions, and the organisms themselves may secrete or contain metal-binding agents. Suppose K_p = stability constant of a metal ion with ligand L_p and K_q = stability constant with ligand L_q. Equating the concentrations of free metal ion in each case we have from Eqn *12.4*

$$\frac{[ML_p]}{[ML_q]} = \frac{K_p}{K_q} \cdot \frac{[L_p]}{[L_q]} \qquad 12.16$$

It follows that, only if the values of K_p and K_q are about equal will the two complexants effectively compete for the metal ion. If $K_p \gg K_q$ and there is an excess of each ligand most of the metal ion will be bound in the L_p complex.

12.11.7 Metal buffers in culture media

The stability constants of some chelating agents of biological importance are given in Table 12.3. Agents suitable for metal buffers in culture media should

TABLE 12.3 Stability constants of some chelating agents of importance in culture media

Chelating agent	Log stability constant							
	Fe^{3+}	Cu^{2+}	Zn^{2+}	Co^{2+}	Fe^{2+}	Mn^{2+}	Ca^{2+}	Mg^{2+}
EDDHA[2]	33·9	15	9·3	—	14·3	—	7·2	2·9
CDTA[1]	27·5	21·3	18·5	18·9	16·3	14·7	12·5	10·3
EDTA[1]	25·1	18·3	16·3	16·2	14·3	13·6	10·7	8·7
NTA[1]	15·9	12·8	10·5	10·6	8·8	7·4	6·4	7·0
Histidine[1]	—	18·3*	12·9*	13·9*	9·3*	7·7*	—	—
8 Hydroxyquinoline	26·3*	25·4*	17·1*	19·5*	15·0*	13·5*	13·2*	12·0*
1:10 Phenanthroline	14·1*	18·0	17·0*	—	21·0*	7·4*	—	—
2:2' Dipyridyl	—	17·9†	13·5†	—	17·6†	6·3†	—	—
SSA	14·1	9·4	—	6·5	—	5·3	—	—
Glycine[1]	—	15·2*	9·5*	8·9*	7·8*	4·7*	—	—
Citric acid[1]	11·4	5·9	5·0	5·0	4·4	3·7	3·6	3·3
Polyphosphate	6·5*	5·5*	6·0*	3·0	3·0	5·5*	3·0	3·2

EDDHA = ethylenedinitrilo-N,N'-bis(2'hydroxyphenyl)-N,N'-diacetic acid
CDTA = 1:2 diaminocyclohexane-N,N-tetra-acetic acid
EDTA = ethylenediamine tetra-acetic acid
NTA = nitrilotri-acetic acid
polyphosphate = $[P_nO_{3n+1}]^{(n+2)-}$ where $n \approx 5$
SSA = 5-sulphosalicylic acid
 The stability constant is $K = [ML]/[M][L]$ where $[ML]$, $[M]$ and $[L]$ are the concentrations of complex, free metal ion and free ligand respectively
* Stability constant is $\beta_2 = [ML_2]/[M][L]^2$, which is obtained from the equilibrium $M + 2L \rightleftharpoons ML_2$
† Stability constant is $\beta_3 = [ML_3]/[M][L]^3$, which is obtained from the equilibrium $M + 3L \rightleftharpoons ML_3$

[1]Bjerrum et al. (1958), Sillén & Martell (1971); [2]Wallace (1962)

not be metabolized. Of the organic ligands, CDTA and EDTA are probably the most immune to metabolic attack. Polyphosphate is heat labile and should be sterilized by filtration. Unfortunately there is little information available about the effect of pH on the stability constants. From Fig. 12.3 it can be seen that at pH 7 the log K_{app} for EDTA is log $K - 3$. In the specificities of the chelating agents given in Table 12.3, on the whole, the affinity is greatest for Fe^{3+} and least for calcium and magnesium. The degree of binding of iron will depend on whether it is present as Fe^{3+} or Fe^{2+} ions. The work of Ratledge (1971) indicates that bacteria take up iron as ferric ions. Those chelating agents which form lipophilic complexes, for example EDDHA and 8-hydroxyquinoline, may permeate the plasma membrane of an organism and assist uptake of the metal ion (Rubin, 1961).

12.12 Design of a culture medium

12.12.1 A minimal medium

The amounts and nature of constituents of culture media are determined by the yields of products and the growth rate required. Estimates of the growth yields can be made from the elementary composition of the biomass, and the energy source requirement may be estimated from knowledge of the ATP yield. Table 12.4 gives the composition of a chemically defined medium (PI), for growth of *Klebsiella aerogenes*, which is designed to give up to 10 g bacterial dry weight/l when excess of oxygen is available. Unless automatic pH control is used, the pH value will fall due to the utilization of ammonia. To avoid adding a large amount of ammonium salt to the medium, ammonia gas may be fed in by pH control as the ammonium ion is consumed. For some organisms, urea or glycine may be used in place of ammonia to reduce the pH change. Special metabolism may require modification of the medium, for instance molybdenum (as molybdate) is required if nitrate is used as the nitrogen source. If large amounts of products other than biomass are to be formed additional substrate for product formation may need to be included. To make any particular nutrient growth-limiting in medium PI, its amount should be decreased until the biomass concentration is significantly less than the maximum attainable.

Enriching the medium with amino acids and nucleic acid bases will spare the glucose, by providing the carbon for assimilation, and increase the growth rate (Hernandez & Johnson, 1967).

In mammalian cell culture the addition of the metabolically inert methylcellulose (0.1%) to cultures in minimal medium greatly reduces the lag before growth (Birch & Pirt, 1970); the mechanism of this important effect is unknown.

12.12.2 Stability of media

The main factors influencing the stability of a medium are: the nature of the constituents; their reactions with each other; temperature, particularly during

TABLE 12.4 Composition of a chemically defined medium (PI) to produce *Klebsiella aerogenes* at concentrations up to 10 g dry biomass/l in aerobic culture

Constituent	Elements provided or other function of constituent	Growth yield assumed (g dry biomass/g element)	Mass of constituent (g)
Water (distilled)§	—	—	1000
Glucose	C, energy	1·10	22·7
NH₄Cl	N	8·75	4·37
KH₂PO₄	P+K	39·1 for P	1·13*
		59·5 for K	
MgSO₄.7H₂O	S+Mg	333 for S	0·232†
		430 for Mg	
CaCl₂.2H₂O	Ca	3·33 × 10³	0·011
FeSO₄.7H₂O	Fe	6·7 × 10³	0·007
MnSO₄.4H₂O	Mn	2·0 × 10⁴	0·002
ZnSO₄.7H₂O	Zn	2·0 × 10⁴	0·002
CuSO₄.5H₂O	Cu	10⁵	0·0004
CoCl₂.6H₂O	Co	10⁵	0·0004
EDTA, disodium salt, dihydrate	chelation	—	0·394‡

The pH should be adjusted to about 7 with NaOH. To increase the buffering power, sodium phosphates (0·1 M in phosphate) are added, e.g. 0·079 M Na_2HPO_4 + 0·021 M NaH_2PO_4 gives pH 7. However, increasing the Na:K ratio may decrease the yield from potassium. The dissolved oxygen tension should be not lower than 15 mmHg to prevent oxygen limitation. The calculated ideal osmolality of the above medium is 313 milliosmolal. With 0·1 M sodium phosphate buffer pH 7 added, the medium is 592 milliosmolal (ideal), which corresponds to a water activity of 0·99.

* The potassium requirement is met by 0·586 g KH_2PO_4
† The amount is the same for sulphur and magnesium
‡ 1 mole EDTA/mole Mg and other non-alkali metals
§ De-ionized water may contain undefined neutral organic compounds

heat sterilization; pH of the medium; oxygen; light. It is convenient to consider the effects of these factors on the main groups of constituents used in media.

The amino acids, tryptophan, glutamine and asparagine, are the most labile amino acids and for this reason should not be sterilized by heat but by filtration. Glutamine is completely decomposed to a γ-pyrrolidone by heating in aqueous solution at 100°C and pH 7 for 3 hours, and even at 37°C it decomposes at an appreciable rate (Griffiths & Pirt, 1967). Cysteine, in the presence of oxygen, is rapidly converted to cystine which has a much lower solubility (about 0·2% at 20°C) than cysteine. However, nutritionally, cysteine and cystine are interchangeable.

Of the water-soluble vitamins, thiamine, riboflavin and pyridoxin are the most prone to decomposition. Thiamine in aqueous solution in the presence of oxygen at 37°C is about 50% oxidized in one week to a biologically inert form. It is labile to autoclaving for 5 min at 121°C (Button, 1969). Riboflavin is destroyed by autoclaving at 121°C for 1 hour at pH 7 but it is more stable at acid pH. Riboflavin is light sensitive and in diffuse room light at 32°C it may be 50% destroyed in 1 hour (Koser, 1968), however, Blaker (1971) found only 13% loss in 157 hours. Also folic acid and pyridoxin are light sensitive, but less so than riboflavin (Blaker, 1971).

Sugars are subject to some decomposition, often accompanied by browning, when they are autoclaved in the presence of inorganic salts and organic compounds. Glycosides with furanoside groups, for example sucrose, at acid pH values hydrolyse on heating, which for many culture purposes is not detrimental, but for defined conditions must be avoided. Generally the sugars in pure solution are stable to autoclaving.

Among the inorganic salts, ammonium salts should be autoclaved at a pH less than 7, otherwise some ammonia is volatilized. In chemically defined media a major loss of magnesium, potassium, ammonium, sodium and phosphate ions can occur through precipitation of the sparingly soluble magnesium ammonium phosphate, magnesium potassium phosphate and magnesium sodium phosphate. The precipitation may not occur until some hours after making up the solution. For this reason magnesium salt should be autoclaved apart from phosphate. Calcium sulphate has a solubility of about 0·2% and the phosphate is sparingly soluble. In media without a chelating agent virtually all iron can be precipitated and made unavailable unless the solution is strongly acid. Seitz filters can absorb ferric ions and create an iron deficiency (Kurowski & Pirt, 1971). Natural media usually contain amino acids and other compounds which chelate trace metals. Many minimal media recommended in the literature are defective because they do not contain a chelating agent which prevents precipitation of iron and other trace metals. Hutner (1972) has recommended preparation of mixtures of trace elements as dry powders. Dry powders of amino acids and vitamins are very convenient for the preparation of tissue culture media.

12.13 Conclusion

Most studies on microbial nutrition have primarily been concerned with the identification of growth factors and choosing the optimal medium based on a qualitative approach, that is adding different substrates in more or less arbitrary amounts. The ultimate aim should be to achieve optimal conditions with defined media and in many processes this remains to be achieved. The use of undefined complex media may mask important nutritional effects, of which a clear example is the stimulation of cephalosporin synthesis by nor-leucine (Drew & Demain, 1973). Development of defined media, which are also optimal for a given process, will depend on elucidation of the functions of all medium constituents and their interactions. Studies on the effects of growth limitation by different nutrients in chemostat culture are proving indispensable to the elucidation of the roles of each nutrient in the culture. These studies have been mostly confined to growth limitation by a few carbon and energy sources, ammonium, potassium, magnesium and phosphate ions, whilst growth factors and trace elements have been neglected. Also studies on substrate-limited growth have been confined mostly to a few bacterial species and need to be extended to many more procaryotes and eucaryotes.

CHAPTER 13

Effects of temperature

13.1 Effects on growth rate

Environmental temperature is a factor to which the biomass is inescapably subject since cell temperature must become equal to the temperature of the culture medium. In contrast, the pH value or the water activity in the cell may not necessarily be equal to that of the outside medium. Temperature affects the rates of cell reactions, the nature of metabolism, the nutritional requirements and the biomass composition.

The effects of temperature on rates of growth are depicted in Fig. 13.1. Over most of the temperature range below the optimum, the temperature coefficient of growth rate corresponds to a Q_{10} of about 2, that is a twofold increase in growth rate per 10°C rise in temperature. The growth rate approaches zero at 10 to 25°C below the optimum temperature. The effects of temperature on duration of growth lag parallel the effects on growth rate (Section 19.5).

13.2 Activation energy of growth

Chemical reaction rates are related to temperature by the Arrhenius equation

$$K = A \, e^{-E/RT} \qquad\qquad 13.1$$

where K is the reaction rate, R is the gas constant, T the absolute temperature, A is a constant dependent on the frequency of formation of activated complexes of the reactants and E is a constant known as the 'activation energy' or 'temperature characteristic'. From Eqn 13.1 we have

$$\log K = \log A - E/2{\cdot}30\,RT \qquad\qquad 13.2$$

Hence a plot of $\log K$ against $1/T$ should be a straight line with slope, $E/2{\cdot}30\,RT$. If we substitute specific growth rate (μ) for the reaction rate (K) in Eqn 13.2, straight line relationships between $\log \mu$ and $1/T$ are obtained over a limited range (Fig. 13.2). Values for the activation energy are given in Table 13.1. Over the short range data of Monod for *Escherichia coli*, and of Trinci for *Aspergillus nidulans*, E is constant. The data of Ingraham for *E. coli* indicate that there is an abrupt increase in the activation energy when the

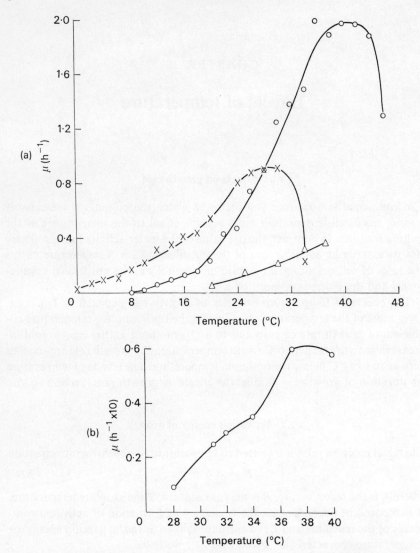

FIG. 13.1 Specific growth rates (μ) of protists and mouse leucocytes as functions of temperature. (a) ○, *Escherichia coli* in rich medium (from data of Ingraham, 1958); X, psychrophilic pseudomonad 21-3c (from data of Ingraham, 1958); △, *Aspergillus nidulans* (from data of Trinci, 1969). (b) Mouse leucocytes (from data of Watanabe & Okada, 1967).

temperature falls below 26°C, and for the psychrophilic pseudomonad when the temperature falls below 12°C. In the upper part of the temperature range the value of E for bacteria is about 14 000 calories and the value roughly doubles in the lower part of the temperature range. The shape of the plot of

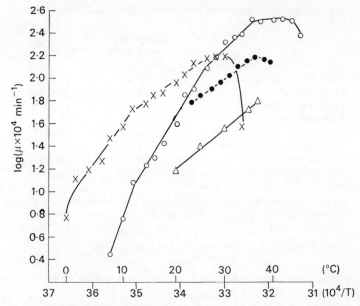

FIG. 13.2 Application of Arrhenius equation to the specific growth rates of microbes. ○, *Escherichia coli* and X, psychrophilic pseudomonad in rich media (from data of Ingraham, 1958); ●, *E. coli* in minimal medium (from data of Monod, 1942); △, *Aspergillus nidulans* (from data of Trinci, 1969).

log μ against $1/T$ for mouse leukemic cells is similar to that for bacteria, however, the value of E is constant only from 37° to 30°C and, below 30°, E increases (Watanabe & Okada, 1967).

It should be noted that the temperature coefficient of the growth rate expressed as Q_{10} is a function of the temperature range since from the Arrhenius equation, it follows that

$$\log Q_{10} = \frac{E}{2 \cdot 30 R} \frac{10}{(T+10)T} \qquad 13.3$$

hence the Q_{10} should vary inversely as the temperature but the effect will be slight over the normal range of temperature for growth. The activation energy is a valuable constant in that it can be used to predict the effect of temperature on growth rate over the normal temperature range.

Changes in the activation energy indicate that differences in the rate-controlling reactions or in metabolic regulation can occur. Evidence for this view is presented by Ng *et al.* (1962) who found that in *E. coli* the repression of β-galactosidase by glucose began to fail below 25°C and disappeared entirely during growth at 10°C.

TABLE 13.1   Values of the activation energy (*E*) for microbial growth

Organism	Temperature range (°C)	*E* (calories)
Aspergillus nidulans[1]	20–37	14 000
Escherichia coli[2]	23–37	13 100
Escherichia coli[3]	26–37	16 200*
Escherichia coli[3]	12–26	28 600
Psychrophilic pseudomonad[1]	12–30	12 600*
Psychrophilic pseudomonad[1]	2–12	23 800
Klebsiella aerogenes[4]	20–40	14 230
Mouse tissue cells [5]	31–38	27 500

[1]From data of Fig. 13.2; [2]from Monod data in Fig. 13.2; [3]from Ingraham data in Fig. 13.2; [4]Topiwala & Sinclair (1971); [5]Watanabe & Okada (1967)

* Re-calculated from data given by Ingraham (1958)

The temperature range for growth of individual species of bacteria extends over about 35°C. Extreme psychrophiles should be capable of growth between −5° and 30°C and extreme thermophiles between 55°C and 90°C. The temperature range for growth of a fungal species extends over about 30°C, and for a mammalian cell type about 12°C.

13.3   Upper limit of growth temperature

The decrease in the growth rate at the upper extreme of the temperature range may reflect either a disruption of metabolic regulation or death of the cells. If death occurs the growth rate of the viable biomass (*x*) is given by

$$dx/dt = (\mu - k)x \qquad\qquad 13.4$$

where μ = specific growth rate and k = specific death rate. The death rate will become dominant at high temperatures if the activation energy for death exceeds that for growth. This was found to be so for *Klebsiella aerogenes* for which the activation energy of death was 32 900 calories compared with a value of 14 230 calories for growth (Topiwala & Sinclair, 1971).

Increase in temperature will eventually cause breakdown of protein structure so that affinity for substrate and enzyme regulators will be affected. The existence of temperature-sensitive enzymes illustrates this point. The thermophilic bacteria, some of which can grow at temperatures up to 90°C, have been shown to possess proteins with exceptional heat resistance (Campbell & Pace, 1968). In contrast, there is no evidence of changed structure in psychrophiles to account for cold resistance.

13.4 Effects on nutrient requirements

Lowering the growth temperature for bacteria can cause a small increase (10 to 20%) in the growth yield from the carbon and energy source. In *Klebsiella aerogenes* the decrease in the growth yield from glucose with increase in temperature was accounted for by increase in the maintenance energy (Topiwala & Sinclair, 1971); the activation energy of the maintenance metabolism was 9000 calories.

The pathway of metabolism of the carbon and energy source can be temperature sensitive. An example is *Lactobacillus brevis* which at 24°C ferments glucose by the heterolactic pathway whereas at 37°C requires fructose as a hydrogen acceptor (forming mannitol) for glucose fermentation (De Ley, 1962).

At a given growth rate, the growth yields from magnesium, potassium and phosphate decrease on lowering the temperature, which reflects the increase in the RNA content of the biomass (Tempest, 1969).

Growth factor requirements can change with temperature. A classical example is *Yersinia (Pasteurella) pestis* which requires more different amino acids and vitamins for growth at 37°C than for growth at 28°C (Hills & Spurr, 1952). Shift up in the growth temperature of *Y. pestis* from 28° to 37°C during growth of the culture must be done gradually otherwise the effect is lethal (Pirt, unpublished). It follows that a number of enzyme syntheses are temperature sensitive in *Y. pestis*. *Escherichia coli* develops requirements for glutamic acid and nicotinamide for growth at 44°C (Ware, 1951). *Saccharomyces cerevisiae* was found to require pantothenic acid and sodium chloride for growth at 38°C but not at 30°C (Beque & Lichstein, 1963).

13.5 Effects on product formation

Secondary metabolism and overproduction of intermediary metabolites may respond to temperature in a different way to growth. In a glucose-limited chemostat culture of *Aspergillus nidulans* the production rate of melanin increases about twofold between 23° and 37°C when the growth rate is maintained constant at 0·05 h^{-1} (Rowley & Pirt, 1972). Overproduction of riboflavin by *Ashbya gossypii* requires growth of the organism at 28°C, although the organism grows equally well at 37°. Resting cells of *A. gossypii* grown at 28°C will overproduce riboflavin at both 28° and 37°, hence it is concluded that growth at the lower temperature causes breakdown of normal regulation of synthesis of the enzyme system which produces riboflavin (Demain, 1972b). The batch culture results of Owen & Johnson (1955) indicate that the optimum temperature for penicillin production is lower than the optimum for growth of the mould, however, to establish this effect definitively will require application of steady-state chemostat culture.

13.6 Effects on microbial composition

Temperature affects the RNA, protein and lipid composition of bacteria and yeasts in the following ways (Hunter & Rose, 1972). At a given growth rate the RNA contents of bacteria or yeasts increase several fold on decreasing the temperature. The total protein content of yeasts may vary up or down with a given change of temperature, depending on the nature of the growth-limiting substrate. Yeast lipids increase their contents of unsaturated fatty acids when the temperature is lowered. In *E. coli*, above the temperature for maximum growth rate, a marked increase in the palmitic acid content occurs at the expense of the unsaturated hexadecenoic and octadecenoic acids (Marr & Ingraham, 1962). Possibly, membrane lipid structure is continuously modified with change in temperature so as to maintain lipid function. Breakdown of membrane structure by the temperature effect might cause loss of metabolites, and stimulate their overproduction.

Antigenic composition of bacteria varies qualitatively and quantitatively with temperature, for example certain antigens of *Y. pestis* necessary for virulence are synthesized during growth at 37° but not at 28°C (Pirt *et al.*, 1961). Such observations suggest that the surface structures of bacteria vary with the growth temperature.

13.7 Mechanisms of temperature effects

The effects of temperature on biomass growth and activity have to be explained basically in terms of the temperature dependence of the structure of cell components especially proteins and lipids, and of the temperature coefficients of reaction rates which in turn depend on the activation energies of the reactions. In response to these primary effects there will probably be many secondary effects on metabolic regulatory mechanisms, specificity of enzyme reactions, cell permeability and cell composition. The secondary effects may, to some extent, be counteracted by special environmental conditions.

CHAPTER 14

Effects of hydrogen ion concentration

14.1 Extracellular and intracellular pH

The influence of hydrogen ions on biological activities is related to either hydrogen ion concentration [H$^+$] or hydrogen ion activity (a_h). The two parameters are proportional so that $a_h = f$[H$^+$] where f is the activity coefficient, which may vary with ionic strength and other factors. The glass electrode responds to hydrogen ion activity so that strictly pH $= -\log (a_h)$ and [H$^+$] can be substituted for a_h only when the activity coefficient is one. In dilute media $f \approx 1$, but this may be far from true when the media are strong salt solutions. As far as cell properties are concerned hydrogen ion activity is the more meaningful parameter so that it is appropriate to express the effects of hydrogen ions in terms of pH.

Plasma membranes are not freely permeable to hydrogen or hydroxyl ions so that intracellular and extracellular hydrogen ion concentrations do not necessarily equilibrate and a gradient of hydrogen ion concentration across the membrane can be expected. According to the chemiosmotic theory (Mitchell, 1973) this gradient of hydrogen concentration together with the membrane electrical potential determine the 'proton motive force' which drives membrane reactions.

14.2 Control of culture pH

In many studies on microbial cultures the pH control has been poor and the pH has not been a constant factor. In order to avoid significant pH drift the limitations of the control methods must be recognized. Culture pH is controlled by either pH buffers or by balancing production and utilization of acids or by automatic addition of acid or base.

The pH buffers are limited both in their pH range and in their capacities for uptake or release of hydrogen ions, also they may have specific physiological effects. Phosphate buffer is universally used for the pH range 6 to 8. For pH values < 6 phthalate has been used and for pH > 8 borate. To keep within the buffer capacity, the acid or base production should be restricted by the design of the medium. Excess calcium carbonate is an effective way of neutralizing large acid production, however, often this method is inconvenient.

It may be possible to balance acid production in the culture by uptake of an anion such as acetate or nitrate which is replaced by hydroxyl ion; base production may be balanced by uptake of cations such as ammonium ions, which are replaced by hydrogen ions. Often uptake of ammonium ion is the sole source of acid production and this may be avoided by substituting urea or glycine for the ammonia.

The limitations of the above methods of pH control are overcome by automatic addition of acid or base dependent on pH measurement by the glass electrode. Usually either hydrochloric or sulphuric acids are added, and one of the bases, sodium, potassium or ammonium hydroxide. Ammonia gas may be preferred to ammonium hydroxide because, the latter even in concentrated (0·880) form can be contaminated with bacterial spores. To study pH effects on a nutrient agar plate, a pH gradient may be set up across the plate by pouring an inclined layer of nutrient agar at one pH value and layering on top nutrient agar of a different pH value (Sacks & Pence, 1958).

14.3 Effects of pH on growth and metabolism

Some effects of pH value on bacterial growth rate are shown in Fig. 14.1. The curves are fairly symmetrical in contrast to the temperature effect curve (Fig. 13.1). Maximum growth rate is maintained over a range of 1 to 2 pH units and the full range for growth is 2 to 5 pH units.

The culture pH has considerable influence on the end products of anaerobic metabolism of carbon and energy sources. *Klebsiella aerogenes* at pH 5 under anaerobic or oxygen-limited conditions produces 2,3-butanediol as the major fermentation product of sugar, whereas at pH 7 much of the butanediol is replaced by lactate. In anaerobic metabolism, many bacteria at acid pH tend to produce neutral products whereas at alkaline pH they tend to produce organic acids. However, there are notable exceptions to this rule, in particular, the lactobacilli and streptococci which at acid pH ferment sugars to lactic acid, and *Acetobacter* which oxidize alcohols to acids at acid pH values. The nature of the end products of sugar fermentations by streptococci and lactobacilli can be pH dependent—at acid pH the product is almost entirely lactic acid—but this gives way to acetic and formic acids when the pH is raised (de Ley, 1962); thus the higher pH favours production of more acid equivalents per mole of substrate. Yeast fermentation of sugar produces ethanol at acid pH whereas, at alkaline pH, glycerol and acetic acid appear. There is evidence that the maintenance energy may be affected by pH value (Section **8.3.3**).

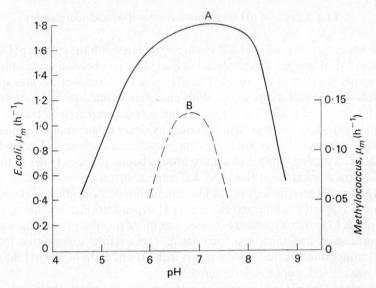

FIG. 14.1   Effects of medium pH value on maximum bacterial growth rates: A,
Escherichia coli in anaerobic casein hydrolysate medium with buffer control of pH
(from data of Gale & Epps, 1942); B, *Methylococcus capsulatus* grown on methane
with pH control by automatic addition of acid or base (Harwood, 1970).

Gale & Epps (1942) observed that in batch cultures of *Escherichia coli*
grown on casein hydrolysate the optimum pH for induction of amino acid
decarboxylases is on the acid side (pH 5), whereas the optimum pH for induction of the deaminases is on the alkaline side (pH 8); hence, the metabolic
changes tended to regulate the medium pH towards neutrality. These results
are complicated by the growth rate changes which accompanied the pH
changes, a difficulty which could now be avoided by chemostat culture.
However, from their observations, Gale & Epps (1942) suggested that the
amounts of the enzymes are regulated so as to maintain a constant activity
and compensate for the effect of medium pH value.

The medium pH, by its effect on the dissociation of an acidic or basic
compound, can influence the inhibitory or toxic action of the compound, for
instance, the toxicity of weak acids such as acetic acid is often pH dependent.
The free acid is more lipid soluble than the ionized form, hence reduction in
pH will favour permeation of the acid into the cell.

The pH-dependence of secondary metabolism may differ from that of
growth and primary metabolism. This was shown for *Aspergillus nidulans*
where the specific rate of melanin formation remains nearly constant at pH
values from 3 to 7 but increases twentyfold between pH 7 and 7·9 (Rowley &
Pirt, 1972).

14.4 Effects of pH on biomass composition and morphology

The antigenic composition of *Yersinia pestis* varies with the culture pH value. The so-called 'antigen 4' is produced only at pH values below 6·9, the optimum being pH 5·9 (Pirt *et al.*, 1961). The pH value of cultures of *Bacillus* species affects the cell wall composition. With phosphate-limited growth, teichuronic acid is produced at alkaline pH values, but is largely replaced by teichoic acid at pH values < 6. When growth is limited by either magnesium or sulphate or ammonium ions, teichuronic acid is not present and teichoic acid is produced with an increasing number of alanine groups bound to the polymer as the pH is decreased towards 5·0 (Ellwood & Tempest, 1972).

The hyphal morphology of the filamentous mould *Penicillium chrysogenum* grown in agitated submerged culture is pH dependent (Pirt & Callow, 1959). At pH 6, long thin hyphae are formed and on raising the pH to 7 the hyphae become short and thick and vacuolated. These results suggest that the cell wall composition of the mould changes with pH and at the higher pH the walls are weaker and more easily ruptured.

In general the pH value during growth considerably affects the nature of the cell surface or envelope materials.

14.5 Molecular basis of pH effects

The culture pH value will affect the medium composition and the nature of the microbial surface through the dissociation of acids and bases. This effect on the cell surface could affect surface properties such as adhesion to glass or metal and flocculation of the biomass. The degree of metal ion chelation will depend on the pH value.

The plasma membrane may to some extent make the cell's interior independent of the medium pH, however, it is clear that metabolism is readily influenced by environmental pH. The equilibration of internal and external hydrogen ion concentrations may be achieved by a weak lipid-soluble acid such as dinitrophenol present in the medium, which can act as a proton conductor across the plasma membrane (Mitchell, 1973). This effect may account for the growth inhibitory properties of the lipophilic acids for some bacteria (Freese *et al.*, 1973). The pH value will affect the conformation of macromolecules which, in the case of enzymes, could affect their K_m values (Dixon, 1953).

It is becoming apparent, largely as a result of the application of automatic instrumental control of pH and chemostat culture, that cell properties can change profoundly over a narrow pH range. However, the molecular basis of the effects is little understood.

CHAPTER 15

Effects of water activity and medium tonicity

15.1 Introduction

Water in living cells has four basic functions. (*i*) As a chemical reactant in the cell, it enters into hydrolyses and condensations. (*ii*) It acts as a solvent for the cell pool of metabolites such as amino acids and their concentrations could be critical in metabolic regulation. (*iii*) It has a mechanical role in maintaining the cell's turgidity due to the hydrostatic pressure which develops in the cell. (*iv*) It has a structural function in hydration of protein and other cell components. This last function depends on the ability of water to form hydrogen bonds with polar groups. Although water is essential for growth, drying the biomass need not be lethal.

The availability of water to the cell is expressed in terms of the osmotic pressure (atmospheres), the tonicity (osmoles) of the medium, or the water activity. In microbe and cell cultures the osmotic pressure has little meaning except in determining whether the cell can resist a given pressure difference across the plasma membrane; tonicity or water activity are more meaningful measures of water availability.

The term ionic strength has been used loosely in microbiology to refer to salt concentration—almost invariably sodium chloride concentration—but it has rarely been used in the physicochemical sense where ionic strength is defined as

$$i = 0.5(m_1 z_1^2 + m_2 z_2^2 + \ldots + m_n z_n^2) \qquad 15.1$$

where m_1, m_2 etc. and z_1, z_2 etc. are the molalites and valencies, respectively, of the ions in the solution. The ionic strength, i, is a measure of 'the intensity of the electrical field due to the ions in a solution' (Glasstone & Lewis, 1964, p. 510); this electrical field influences the activities of both the ions and molecules present in the solution. The present significance of i is summed up in Section 15.10.

15.2 Definition of water activity

The activity of water in a solution may be expressed in terms of the water vapour pressure in the same way as the activity of a gas such as oxygen in

147

solution is expressed as the equivalent pressure of the gas. However, it has been found more convenient to relate the water activity of a solution (a_w) to that of pure water. Accordingly the water activity of the solution is expressed as the ratio of the water vapour pressure of the solution (p_s) to that of pure water (p_w) at the same temperature, thus

$$a_w = p_s/p_w \qquad\qquad 15.2$$

It follows that a_w is equivalent to the relative humidity of the atmosphere in equilibrium with the solution.

15.3 Relation of water activity to solute concentration

According to Raoult's law for a solution of an ideal non-electrolyte, the relative lowering of the solvent vapour pressure is equal to the molar fraction of the solute in the solution, that is

$$(p_w - p_s)/p_w = n/(n+N) \qquad\qquad 15.3$$

where n = moles of solute and N = moles of water in the solution. From Eqn 15.2 and 15.3 it follows that

$$a_w = 1 - n/(n+N) \qquad\qquad 15.4$$

or

$$a_w = N/(n+N) \qquad\qquad 15.5$$

If the solute is an electrolyte, instead of n we put vn where v is the number of ions per molecule. If the solute does not behave ideally the real number of particles of solute will appear to be ϕvn where ϕ is the *osmotic coefficient* (Section **15.4**).

15.4 Relation of water activity to osmotic pressure

According to the Van't Hoff equation the osmotic pressure π of a solution of an ideal non-electrolyte is given by

$$\pi V = nRT \qquad\qquad 15.6$$

where V is the volume of the solution, n is the number of moles of solute present, R is the gas constant and T the absolute temperature. A more accurate equation is that of Morse,

$$\pi V' = nRT \qquad\qquad 15.7$$

where V' is the volume of the solvent. If the solute is ionized we substitute vn for n where v is the number of ions from each molecule of solute. For a nonideal solute the osmotic pressure is corrected by the osmotic coefficient ϕ, that is

$$\pi V' = \phi v n RT \qquad\qquad 15.8$$

The departure of a solute from ideal behaviour is the result of differences in the degrees of association, due to secondary valencies, between molecules of solvent, molecules of solute and solvent and solute molecules. Association between solute molecules makes $\phi < 1$, and association between solute and water makes $\phi > 1$.

If in Eqn 15.7, V' is expressed in litres of water then n/V' can be equated with molality (although strictly the molality is mol/kg water) and Eqn. 15.7 becomes

$$\pi = mRT \qquad\qquad 15.9$$

where m = molality. If $m = 1$, $R = 0.0821$ litre.atm deg^{-1} mol^{-1}, at 0°C we find $\pi = 22.4$ atm. It should be noted that π is directly proportional to the absolute temperature, so that at 30°C a 1 osmolal solution will exert an osmotic pressure of 24.85 atm.

The relation between water activity and osmotic pressure derived from thermodynamics (Section 15.11) is

$$\pi = -\frac{RT}{\bar{V}} \ln a_w \qquad\qquad 15.10$$

where $\bar{V}$ is the volume of 1 mol water. Substituting for π from Eqn 15.9 we obtain

$$m_0 = v m \phi = -(1/\bar{V}) \ln a_w \qquad\qquad 15.11$$

where m_0 is the real osmolality. Changing to logarithms to base 10 and putting $\bar{V} = 0.018$ l/mol we find

$$m_0 = -128 \log a_w \qquad\qquad 15.12$$

15.5 Measurement of tonicity and water activity

There are four convenient methods for the measurement of the tonicity (osmolality) or water activity of a culture medium.

1.   The osmotic coefficients of many common salts and other medium constituents can be found in physical tables (Robinson & Stokes, 1959) and from these values the real osmolalities of the constituents can be calculated.

2.   The depression of the freezing point of a solution (ΔT) is proportional to the real osmolality, that is $m_0 = K \Delta T$; the value of the constant K is found in physical tables. This method is applicable provided that no crystallization of a medium constituent occurs before the freezing point is reached.

3. Isopiestic methods depend on determining whether water vapour tends to pass from the solution to that of a standard solution of known tonicity (Cejkova, 1965; Scott, 1957).
4. In the dew point method the atmosphere in a dew point apparatus is equilibrated with a sample of medium. From the dew point the water vapour pressure of the solution can be calculated. This method has been developed to estimate water activity to within ± 0.003 (Anagnostopoulos, 1973).

The dew point method is most suitable for the higher tonicities whereas the freezing point method is the more accurate with dilute solutions.

15.6 Use of terms, tonicity and water activity

The water activity parameter has proved useful in studies on concentrated media and especially in studies on microbial spoilage which is prevented by low water activity. In such systems also relative humidity, which can be equated with water activity, is meaningful. In the dilute culture media used for many microbes and tissue cells, where the water activity varies little from unity, it is more convenient to express the water relation in terms of the tonicity expressed as real osmolality.

15.7 Tonicity of cell contents

Studies on the swelling of bacterial protoplasts in media of different tonicities show that the protoplasts behave as osmometers (Gilby & Few, 1959). Whole cells of *Escherichia coli* swell or contract according to the tonicity of the medium and the phenomenon has been applied to study of solute permeation into bacteria (Mitchell & Moyle, 1956). Thus it appears that the cell walls of *E. coli* are elastic; in contrast, micrococci appear to have rigid cell walls incapable of allowing the cell to change its volume. On the whole, these results suggest that the intracellular and extracellular water activities equilibrate.

The internal tonicity of *E. coli* has been found to be 0·6 osmolal and that of a *Micrococcus* species 1·0 osmolal (Mitchell & Moyle, 1956); the corresponding osmotic pressures at 30°C are 15 atm and 25 atm. The higher value for the micrococcus accords with its ability to grow at much higher medium tonicities than *E. coli*. The extent to which the internal tonicity is affected by the growth conditions has not been investigated.

15.8 Relation of growth rates to tonicity and water activity

The optimum tonicity of culture media has been given little attention and is commonly assumed to be about the same as that of mammalian physiological saline, a view which reflects the medical origins of microbiology. Data on the influence of water activity and medium tonicity on the growth rates of microbes, given in Fig. 15.1 and Table 15.1, indicate that the optimum tonicity

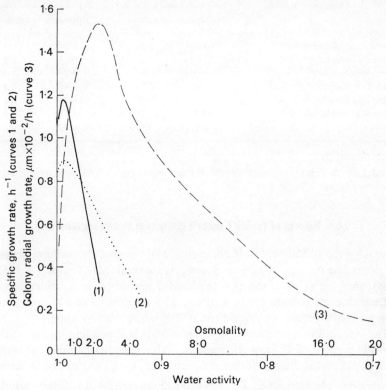

FIG. 15.1   Specific growth rates as functions of water activity and tonicity of the medium. Curve 1, *Salmonella newport* from data of Christian & Scott (1953); curve 2, *Staphylococcus aureus* from data of Scott (1953); curve 3, *Aspergillus amstelodami* from data of Maguire & Scott quoted by Scott (1957) (redrawn with permission of Academic Press). Real osmolalities calculated from, $m_0 = -128 \log a_w$.

for growth rate may differ considerably from that of physiological saline (0·9% sodium chloride). The mould *Aspergillus amstelodami* is termed xerophilic since it is capable of growth at a very low water activity. The microbes most tolerant of low water activities are the osmophilic yeasts for growth of which the minimum water activity is 0·6. The optimum water activity for the growth rate of *A. amstelodami* is independent of whether sucrose, glucose, magnesium chloride, sodium chloride or glycerol are used to control the tonicity, although the maximum value of the growth rate is dependent on the nature of the solute and decreases with the solutes in the order given above, sucrose first (Scott, 1957).

TABLE 15.1 Optimal and minimal water activities and corresponding tonicities required for growth of some organisms

Organism	Water activity		Tonicity (m_0*)	
	optimum	minimum	optimum	maximum
Pseudomonas fluorescens[1]	0·999	0·96	0·05	2·3
Klebsiella aerogenes[2]	0·999	0·95	0·05	2·9
Salmonella newport[3]	0·994	0·95	0·28	2·9
Staphylococcus aureus[3]	0·994	0·86	0·28	8·4
Aspergillus niger[3]	0·975	0·86	1·28	8·4
Aspergillus amstelodami[3]	0·96	0·70	2·15	19·8
Mouse L cells[4]	0·995	0·993	0·30	0·40

* Real osmolality

[1]Wodzinski & Frazier (1960), [2]Wodzinski & Frazier (1961), [3]Scott (1957), [4]Pirt & Thackeray (1964)

15.9 Effects of tonicity on cell composition and metabolism

Increasing the sodium chloride composition of the growth medium increases the potassium content of *Klebsiella aerogenes* several fold (Tempest & Meers, 1968); thus the value of the growth yield for potassium is dependent on the sodium chloride content of the medium. This appears to be a tonicity rather than a specific sodium chloride effect (Christian & Waltho, 1964).

The cell wall composition of *Bacillus subtilis* is dependent on the sodium chloride content of the growth medium. With phosphate-limited growth at neutral pH and sodium chloride concentration < 1%, the cell walls contain teichuronic acid as a major polymer, but on increasing the sodium chloride content of the medium from 1% to 6%, teichuronic acid is replaced by teichoic acid (Ellwood, 1971).

Decreasing the water activity of the growth medium for *Staphylococcus aureus* from 0·993 to 0·90, by means of sodium chloride, decreases the cell water content by about 50% and concomitantly increases the concentration of sodium chloride and the amino acid pool in the cell (Christian & Waltho, 1964). Also a high sodium chloride concentration in the medium greatly increases the L-proline concentration in the amino acid pool of some bacteria (Christian & Hall, 1972). The halophilic bacteria (Larsen, 1967) require a high sodium chloride concentration for growth (5 to 25%). This requirement is not just for a high tonicity but specifically for a high concentration of alkali metal chloride. These bacteria have unusually high contents of potassium, sodium and chloride ions and their enzymes are unusual in that they require high sodium or potassium chloride concentrations in order to function.

A high sodium chloride concentration increases the maintenance energy of yeast (Section **8.3.3**) and changes the anaerobic metabolism of glucose to yield glycerol instead of ethanol (Watson, 1970; Tajima & Yoshizumi, 1972).

15.10 Mechanisms of tonicity effects

The effects of changes in solute concentration of media, if they are non-specific depend either on water activity or on ionic strength, i, as defined in Eqn *15.1*. Generally the extent to which these different factors are involved in the changes is unknown. A preliminary study (Parkes & Pirt, unpublished) of the effects of tonicity changes with different electrolytes and non-electrolytes added to growth media for *Klebsiella aerogenes*, showed no correlation between growth rate and ionic strength although the growth rate was correlated with the water activity in a manner similar to that shown by *Salmonella* (Fig. 15.1).

There is no evidence to suggest that the turgidity of the cell *per se* is important in growth, although a possible adverse effect of a hypotonic medium is cell rupture. This is more likely to happen with organisms such as mycoplasmas or mammalian cells which have no cell walls to protect the plasma membrane, or if the cell walls are weakened. The concentration or dilution of cell metabolite pools by the osmotic effect of tonicity changes could affect regulation of metabolism by feedback and repression mechanisms. Dehydration could affect the conformation of macromolecules and thereby affect their function.

There is no evidence that, at a given pH value, the electric field effect of ionic strength, i, is of any importance apart from its effect on water activity.

The ability of ions to associate with water molecules and so decrease the water activity is given by the Hofmeister series, that is, for anions: sulphate > acetate > chloride > nitrate; and for cations: magnesium > calcium > lithium > sodium > potassium; however, the anions have a stronger effect than the cations (Glasstone & Lewis, 1964). Therefore, to minimize the effect of ionic strength on water activity in culture media, one should use electrolytes with those ions lowest in the Hofmeister series. Some support for this view is given by the finding that the respiration rate of non-growing bacteria is decreased by cations and the effect becomes more marked on going up the Hofmeister series from potassium (Ingram, 1947).

15.11 Derivation of the relation between osmotic pressure and water activity

This relation can be derived by equating the net work required to remove 1 mol of water from the solution by osmosis with that required to remove it by evaporation and condensation. Suppose that, by means of the device shown in Fig. 15.2a, under isothermal and reversible conditions 1 mol of water is expelled from the solution through the semi-permeable membrane by applying a pressure slightly greater than the osmotic pressure of the solution (π). The work done will be $\pi \bar{V}$ where $\bar{V} =$ volume of 1 mol water in the solution.

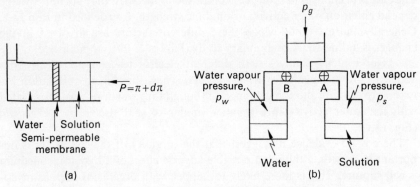

FIG. 15.2   (a) Device to remove water reversibly from a solution by osmosis: $\pi =$ osmotic pressure of solution, $P =$ hydrostatic pressure applied to piston. (b) Device to remove water reversibly from solution by evaporation and condensation: A and B are valves; p_g is pressure applied to piston; p_s and p_w are the water vapour pressures of solution and water respectively.

Evaporation and condensation are supposed to be reversibly and isothermally carried out by means of the device depicted in Fig. 15.2b. First with valve A open and B closed 1 mol of water in the solution is evaporated by making the pressure on the piston slightly less than the water vapour pressure, p_s. The work done during the evaporation will be $p_s V_1 = RT$, where $V_1 =$ volume of water vapour at pressure p_s. With valves A and B closed the vapour is then compressed from pressure p_s to p_w and the work done on 1 mol will be $RT \ln (p_w/p_s)$. Then valve B is opened and the vapour is condensed into pure water which involves doing on the system an amount of work $p_w V_2 = RT$ where $V_2 =$ volume of 1 mol water vapour at pressure, p_w. The work done in evaporation and condensation is in opposite senses and cancels out. Thus the net work done is $RT \ln (p_w/p_s)$. Since the net work done by the osmotic and vapour mechanisms must be the same we have

$$\pi \bar{V} = RT \ln (p_w/p_s) \qquad\qquad 15.13$$

and hence

$$\pi = -\frac{RT}{\bar{V}} \ln a_w \qquad\qquad 15.14$$

This expression can be derived also from Eqn 15.4 if we make the approximation $n/(n+N)=n/N$, which is valid for dilute solutions; then for an ideal non-electrolyte we have from Eqn 15.4

$$a_w = 1 - n/N \qquad\qquad 15.15$$

Putting $y=n/N$ and taking logarithms we have $\ln a_w = \ln (1-y)$. From the expansion of $\ln (1-y) = -y - y^2/2 - y^3/3\ldots$, it follows that for small values of y, $\ln (1-y) \approx -y$, and with this substitution Eqn 15.15 becomes

$$\ln a_w = -n/N \qquad\qquad 15.16$$

If m is the molality of the solution, $n/N = m/55 \cdot 5$, since 1 kg water $= 55 \cdot 5$ mol. Substituting in Eqn 15.16 and changing to logarithms to base 10 we have

$$m = -128 \log a_w \qquad\qquad 15.17$$

which has been shown to follow from Eqn 15.14 (Section **15.4**). Since an approximation was made in deriving Eqn 15.16, Eqn 15.5 and 15.17 do not give identical values for water activity, but the discrepancies are negligible for water activities $> 0 \cdot 9$.

CHAPTER 16

Product formation in microbial cultures

16.1 Introduction

The development of a microbial process for the formation of products such as antibiotics, amino acids or protein is aimed at maximizing three things: the yield of product per gram of substrate, the concentration of product and the rate of product formation. The main features of the development of a process are specified in Table 16.1. Not all of the factors indicated in Table 16.1 are of importance in every process, however, with a new process, *a priori* none of the factors should be disregarded entirely. All four aspects of process development given in Table 16.1 are basically concerned with adjustment of metabolic

TABLE 16.1   Features of fermentation process development

I	Initial selection of strain of organism
II	Determination of optimum values of temperature, pH value, tonicity and oxygen supply
III	Determination of optimum nutritional regimen and biomass concentration
IV	Modification of genetic structure of the organism to increase the product formation

regulation in the organism. Normally the metabolism is adjusted to produce only the minimum amount of essential metabolites, together possibly with small amounts of non-essential secondary metabolites. This usually means that practically all of the carbon source is converted into biomass and the end products of energy metabolism. Successful development of product formation entails interfering with the metabolic regulation so as to make the organism overproduce the desired product. The means for interfering with metabolic regulation applied to fermentation processes have been discussed by Demain (1972b). Unfortunately, knowledge of the pathways of biosynthesis of products is often insufficient to make possible deliberate modification of the control mechanisms at selected points. Therefore, instead, one has to resort to screening, at random, different environmental conditions and mutated organisms for improved productivity. The most scope for deliberate alteration of the metabolic regulation has been found in production of amino acids and nucleotides. The possibility of inhibition of product synthesis by the product itself has been alluded to by Demain (1972b, p. 351). There has been little evidence for it with secondary metabolites but the observation of inhibition of the penicillin fermentation by high concentrations of penicillin (Gordee & Day, 1972) affords a striking example.

Fermentation products are classified in Table 16.2 on the basis of their

TABLE 16.2   Classification of fermentation products

	Class	Examples
I	End products of energy metabolism	Ethanol, methane
II	Energy storage compounds	Glycogen
III	Enzymes	
	extracellular	Amylases
	intracellular	β-Galactosidase
IV	Structural components of cells	Single cell protein, antigens
V	Intermediary metabolites	Vitamin B_{12}, amino acids, citric acid
VI	Secondary metabolites	Antibiotics
VII	Transformed substrates	Steroids
VIII	Viruses	Poliomyelitis

relation to the structure and function of the organism. The processes in each class have some control principles in common, for instance energy storage compounds are normally formed only when the energy source is not growth-limiting (Wilkinson & Munro, 1967). The products may be excreted into the medium or retained in or on the biomass. The exocellular products may be soluble or insoluble and extreme overproduction may lead to precipitation of the product, for instance in oxytetracycline production.

The term *secondary metabolites* refers to non-essential metabolites, examples of which are the antibiotics and the giberellins. Since secondary metabolism is not essential for growth it is possible for non-producing variants of the organism to be selected during long continuous growth.

It is remarkable that no commercial microbial product except protein for feedstuffs and, to a small extent, beer is produced by chemostat culture. This reflects the highly empirical nature of the original batch processes and lack of knowledge of the crucial controlling conditions.

16.2 Relation of growth rate to product formation rate

16.2.1 Growth-linked product

In general the rate of product formation is given by

$$dp/dt = q_p x \hspace{3cm} 16.1$$

where p = product concentration, x = biomass concentration and q_p is the specific rate of product formation. When the product is 'growth-linked' or 'growth-associated' the amount of product formed is directly proportional to the biomass formed. Hence we have

$$dp = Y_{p/x} \, dx \qquad\qquad 16.2$$

where $Y_{p/x}$ is the product yield referred to biomass formed. It follows that

$$dp/dt = Y_{p/x} \, dx/dt = Y_{p/x} \mu x \qquad\qquad 16.3$$

If we express the product yield in terms of the substrate used, we have

$$dp = Y_{p/s} \, ds \qquad\qquad 16.4$$

where $Y_{p/s}$ is the product yield referred to substrate utilized. Hence we obtain

$$dp/dt = Y_{p/s} \, ds/dt = Y_{p/s} \mu x / Y_{x/s} \qquad\qquad 16.5$$

where $Y_{x/s}$ is the biomass or growth yield referred to the substrate utilized. Comparison of Eqn *16.5* and *16.3* shows that

$$Y_{p/s} / Y_{x/s} = Y_{p/x} \qquad\qquad 16.6$$

The specific rate of product formation, it follows from Eqn *16.1* and *16.3*, is given by

$$q_p = Y_{p/x} \mu \qquad\qquad 16.7$$

Growth-linked processes are generally those which are essential to the function of the organism. They include cell components such as cell walls and essential enzymes. The product yield $Y_{p/x}$ may be proportional to the specific growth rate, one example is RNA production. An unusual example, is the enzyme glucoside 3-dehydrogenase of *Agrobacterium tumefaciens*, which transforms sucrose to 3-ketosucrose (Kurowski, 1974), in which case $Y_{p/x} = k\mu$, where k is a constant, hence

$$q_p = Y_{p/x} \mu = k\mu^2 \qquad\qquad 16.8$$

16.2.2 Non-growth-linked product

Non-growth-linked product formation can be of two types: either the q_p value is independent of growth rate or it varies with specific growth rate in a complex way. An illustration of the first case is the production of penicillin by *Penicillium* which is independent of specific growth rate at values greater than about 0.015 h^{-1}. At lower values of the growth rate, decay of the biosynthetic activity occurs (Pirt & Righelato, 1967).

The q_p of a non-growth-linked product can be a complex function of the specific growth rate. An example of this type is melanin formation by *Aspergillus niger* (Rowley & Pirt, 1972), which is represented by

$$q_p = q_p^{\max} - k\mu \qquad\qquad 16.9$$

where $q_p^{\max}$ and k are constants. The formation of cyclodextrin from starch by *Bacillus macerans* (Lane & Pirt, 1973) and spore production by *Bacillus subtilis* (Dawes & Thornley, 1970) are similar.

When product formation is partly growth-linked and partly independent of growth rate we have

$$q_p = Y_{p/x}\mu + \beta \qquad\qquad 16.10$$

Formation of end products of energy metabolism follow this relation where β includes the product formation which results from either the maintenance energy requirement or uncoupling of ATP production. Lactic acid production from sugar by *Lactobacillus* species follows this model (Luedeking & Piret, 1959).

16.3 Product decomposition rate

The product accumulation rate will be influenced by the product decomposition rate so that the general expression for product accumulation rate in a batch culture will be

$$dp/dt = Y_{p/x}\mu x + \beta x - Z \qquad\qquad 16.11$$

where the first two terms on the right-hand side of Eqn *16.11* represent the growth-linked and non-growth-linked product formation respectively and Z is the product decomposition rate. The function Z could be proportional to the biomass concentration if the decomposition is enzymic. Decomposition of energy storage compounds can occur rapidly when the energy source is exhausted, for example glycogen in yeast (Chester, 1963). Penicillin has a significant spontaneous decomposition rate. There is also evidence for enzymic decomposition of penicillin during the fermentation (Gordee & Day, 1972).

16.4 Product formation in batch culture

16.4.1 During growth

The product formed during growth of a batch culture in an infinitely small time interval dt is given by

$$dp = q_p x\, dt \qquad\qquad 16.12$$

where x is the biomass concentration (see Fig. 16.1). If q_p is constant, the product concentration after time t, is given by

$$\int_{p_0}^{p} dp = q_p \int_{0}^{t_1} x \, dt \qquad\qquad 16.13$$

that is

$$p = p_0 + q_p \int_{0}^{t_1} x \, dt \qquad\qquad 16.14$$

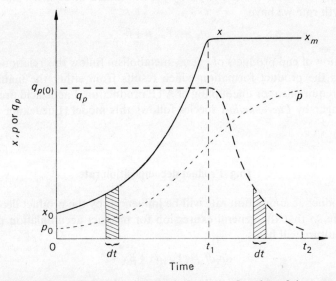

FIG. 16.1 Product accumulation in a batch culture as a function of time. *Symbols:* x = biomass concentration, p = product concentration, q_p = specific rate of product formation.

The value of the integral in Eqn *16.14* may be evaluated graphically from the area under the x, t curve. Alternatively we may integrate the expression for x. If growth is exponential we may write $x = x_0 e^{\mu t}$ in Eqn *16.14*, hence we find

$$p = p_0 + q_p x_0(e^{\mu t_1} - 1)/\mu \qquad\qquad 16.15$$

The increase in biomass may be made linear if the growth-limiting substrate is fed at a constant rate by diffusion capsule (Section **21.6**).

16.4.2 Decay of synthetic activity

After growth ceases, the rate of product synthesis eventually decays. The decay probably begins when the growth rate falls to a critical value close to zero. If it is assumed that q_p is constant until growth ceases at time t, and that subsequently the q_p decreases, as shown in Fig. 16.1, then during the phase of decay of the synthetic activity, the production in a very small time interval dt will be

$$dp = x_m q_p \, dt \qquad\qquad 16.16$$

where x_m is the maximum biomass concentration. The increase in product concentration is given by

$$\int_{p_1}^{p_2} dp = x_m \int_{t_1}^{t_2} q_p \, dt \qquad\qquad 16.17$$

where p_1 and p_2 are the product concentrations at times t_1 and t_2 respectively (Fig. 16.1). The integral of $q_p \, dt$ is the area under the q_p and t curve from t_1 to

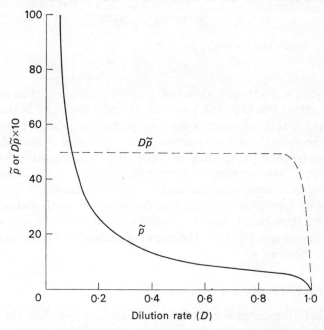

FIG. 16.2 Non-growth-linked product formation in a chemostat culture when q_p is independent of growth rate. *Parameters:* concentration of substrate in feed medium $s_r = 10$, $K_s = 0.01$, $Y_{x/s} = 0.5$, $\mu_m = 1.0$, $q_p = 1.0$, $\tilde{p} =$ steady-state product concentration, $D\tilde{p} =$ output rate for product.

t_2. If decay of the enzyme activity is exponential, or the enzyme decay process is a first order reaction so that $dq_p/dt = K q_p$ where K is a constant, then

$$q_p = q_{p(0)} \, e^{-Kt} \qquad\qquad 16.18$$

hence

$$\log q_p = \log q_{p(0)} - \frac{K}{2.30} t \qquad\qquad 16.19$$

and the slope of the graph gives the decay rate constant, K; the half life of the activity will be $(\ln 2)/K$. Exponential decay of the synthetic activity was found to occur for the glucoside 3-dehydrogenase system in *Agrobacterium* (Fensom & Pirt, 1972) with minimum half life of about 8 hours. In the case of penicillin synthesis the decay was found to be approximately linear (Pirt & Righelato, 1967) rather than exponential.

16.5 Product formation in a chemostat culture

16.5.1 General features

The product formation rate in a chemostat culture is given by

$$dp/dt = q_p x - Dp \qquad 16.20$$

and in the steady state when $dp/dt = 0$

$$\tilde{p} = q_p \tilde{x}/D \qquad 16.21$$

If the product is strictly growth-linked, it follows that the product concentration and output rate $(D\tilde{p})$ will vary with D in the same way as the biomass does. If the q_p is independent of the growth rate, the product concentration varies inversely as the dilution rate and over a wide range of dilution rates the output rate is constant (Fig. 16.2). However, as $D \to 0$, eventually the assumption that q_p is constant becomes invalid because either decay of the enzyme activity begins or some required substrate will be exhausted.

If product formation is partly growth-linked and partly independent of growth rate (Eqn *16.10*), then the product concentration will vary with dilution rate as shown in Fig. 16.3. The output rate when $D=0$ is the non-growth-linked contribution, $\beta \tilde{x} = \beta Y_{x/s} S_r$.

16.5.2 Transient production

If a product is formed transiently in a chemostat culture as shown in Fig. 16.4, the cumulative output of product (p_c) from the initiation of product formation is given by

$$p_c = D \int_0^{t_1} p \, dt \qquad 16.22$$

The integral $p \, dt$ will be the area under the p, t curve (Fig. 16.4) between time zero and t_1. This method was used to evaluate the cumulative production of interferon in a chemostat culture (Tovey *et al.*, 1973).

16.5.3 Effect of biomass concentration

One way of increasing the product formation rate is to increase the biomass

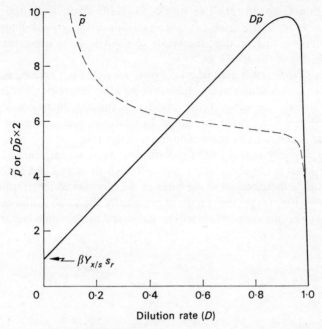

FIG. 16.3 Concentration ($\tilde{p}$) and output rate ($D\tilde{p}$) of product from chemostat culture when product formation is partly growth-linked, that is $q_p = Y_p \mu + \beta$. *Parameters:* concentration of substrate in medium feed $s_r = 10$, $K_s = 0.01$, $Y_{x/s} = 0.5$, $Y_{p/x} = 1.0$, $\beta = 0.1$.

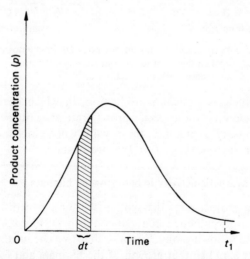

FIG. 16.4 Transient formation of product in a chemostat culture with constant biomass and constant dilution rate.

concentration (Eqn *16.11*). The means available to increase the biomass concentration are to increase the concentration of the growth-limiting substrate, to remove any product inhibitory to growth, or to use some form of biomass feedback (Section **6.3**).

If biomass feedback is used when the product yield ($Y_{p/x}$) is a constant, and the product is exocellular and soluble, the product concentration cannot be increased beyond the value $Y_{p/s}s_r$, where s_r is the concentration of growth-limiting substrate in the feed medium. In this case, increase in the product output rate is achieved by increasing the dilution rate.

When q_p is independent of the growth rate, the concentration of an exocellular product in the chemostat with feedback will exceed that in the simple chemostat by a factor equal to the ratio of the biomass concentrations in the two systems (Fig. 16.5). This factor is practically the same as the concentration factor; also the output rate will be increased by the same factor.

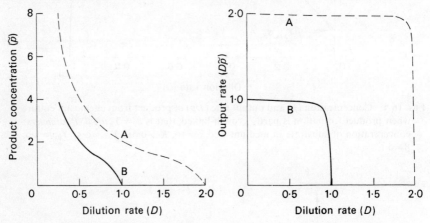

FIG. 16.5   Product concentration, $\tilde{p}$ and output rate, $D\tilde{p}$ in chemostat culture with biomass feedback (broken line A) and without feedback (continuous line B) when q_p is constant; biomass 'concentration' factor = 2.

In processes such as penicillin production, in which decay of synthetic activity begins below a certain specific growth rate, the critical dilution rate in a chemostat necessary to maintain the q_p value will be increased by the 'concentration factor' (α) since $\mu = \alpha D$ in the steady state.

16.5.4 Decay of synthetic activity in non-growing biomass

At low and zero growth rates the synthesis of non-growth-linked products decays. The decay of synthetic activity in non-growing biomass in a chemostat is modelled in the following way. Suppose that two chemostats are arranged in series (Section **6.4.1**) so that growth of the biomass and formation of the desired enzyme activity occurs in the first stage, and product formation by non-growing biomass occurs in the second stage. It is assumed that the second stage is fed with all the necessary substrates for the desired synthesis. Also, it is supposed that growth of the biomass in the second stage is prevented by exhaustion of some substrate but, if this is the energy source, then the maintenance ration is supplied so that the biomass remains constant. We assume that the synthetic activity of the biomass decays according to some function of time as shown in Fig. 16.6. Let df be the fraction of the biomass with resi-

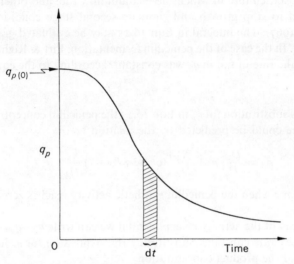

FIG. 16.6 Decay of synthetic activity (q_p) in an element of biomass as a function of time.

dence times between t and $t+dt$, then the contribution to the production rate by the biomass with residence time between t and dt will be

$$(dp/dt)_i = q_p x \, df \qquad 16.23$$

where x is the concentration of biomass. From the distribution of residence times for the chemostat we can substitute $df = D \, e^{-Dt} \, dt$ (Section 5.4). In the steady state the total production rate (dp/dt) will be the sum of all the individual contributions between residence times from 0 to τ, where τ is the time when $q_p = 0$, so

$$\sum (dp/dt)_i = dp/dt = xD \int_0^\tau q_p \, e^{-Dt} \, dt \qquad 16.24$$

Let D_{12} be the partial dilution rate from the first stage to the second stage of the process, then the output rate of product for the second stage in the steady state is given by the balance

$$\text{output} = \text{input} + \text{production}$$

that is

$$D_2\tilde{p}_2 = D_{12}\tilde{p}_1 + D_2\tilde{x}_2 \int_0^\tau q_p\, e^{-D_2 t}\, dt \qquad\qquad 16.25$$

where $\tilde{p}_1$ and $\tilde{p}_2$ are the product concentrations in the first and second stages respectively, and $\tilde{x}_2$ is the biomass concentration in the second stage. The value of q_p as a function of t can be determined experimentally from a single stage chemostat culture in which the medium flow rate and other conditions are changed to stop growth and simulate second stage conditions (Pirt & Righelato, 1967). The integral in Eqn 16.25 may be evaluated graphically or analytically. In the case of the penicillin fermentation, Pirt & Righelato (1967) found that the rate of fall in q_p was constant, according to the equation

$$q_p = q_{p(0)} - k_p t \qquad\qquad 16.26$$

Making this substitution for q_p in Eqn 16.25 the penicillin concentration in the second stage could be predicted by the relation

$$\tilde{p}_2 = \frac{D_{12}}{D_2}\tilde{p}_1 + \frac{\tilde{x}_2}{D_2}\left\{(e^{-D_2 q_{p(0)}/k_p} - 1)\frac{k_p}{D_2} + q_{p(0)}\right\} \qquad\qquad 16.27$$

where the time when the penicillin synthetic activity reaches zero is given by $q_{p(0)}/k_p$.

If the decay of the activity be exponential we can write $q_p = q_{p(0)}\, e^{-Kt}$ where K is the decay rate constant. Substituting this expression for q_p in Eqn 16.25 we obtain for the product concentration

$$\tilde{p}_2 = \frac{D_{12}}{D_2}\tilde{p}_1 + \tilde{x}_2 q_{p(0)}/(K + D_2) \qquad\qquad 16.28$$

16.6 Control of decay of synthetic activity

The problem of maintaining non-growth-linked synthetic activity is of prime importance in secondary metabolite fermentations such as penicillin production. Four basic causes of the decay can be specified: (*i*) inhibition by metabolites or products; (*ii*) loss of intermediary metabolites by diffusion out of the biomass; (*iii*) reversible enzymic inactivation of an enzyme as described for glutamine synthetase (Holzer & Düntze, 1971); (*iv*) irreversible decomposition of an essential enzyme. Particular importance is attached to cause (*iv*) (Thurston, 1972).

In *Penicillium chrysogenum* used for penicillin production the decay process occurs at growth rates below $0.014\ h^{-1}$ (Pirt & Righelato, 1967) and the rate of decay, if growth is stopped, varies inversely as the previous growth rate. In contrast, the loss of the 3-keto dehydrogenase activity from *Agrobacterium* in non-growing cells does not depend on the previous growth rate (Kurowski *et al.*, 1973). In the latter case, the growth-limiting substrate had some influence on the decay rate; it was least with carbon-limited growth and greatest with phosphate-limited growth.

Enzyme decomposition through turnover of protein is inherent in nongrowing cells so that inhibitors of protein turnover may also minimize enzyme decay. Changes in biomass properties as the growth rate approaches zero are discussed in Chapter 18.

16.7 Effects of environmental conditions on formation of microbial products

16.7.1 General

The influence of environmental factors on product formation in fermentations has, with few exceptions, been expressed in terms of the final or maximum product concentration. However, this overall effect is determined by the following independent factors: (*i*) biomass concentration; (*ii*) the specific production rate, q_p; (*iii*) the product yield from the substrate, $Y_{p/s}$; (*iv*) the duration of the synthetic activity; and (*v*) the rate of decomposition of the product. Ultimately, optimization must be analysed in terms of these five factors. The biomass specific growth rate is important largely for its effect on q_p.

The conditions optimum for growth and induction of the synthetic activity may be different from the optimum conditions for product formation. A case in point is riboflavin production (Section 13.5).

A great variety of effects are exploited to increase product formation. The effects of the physicochemical factors, temperature, pH value, tonicity and dissolved oxygen tension are discussed in the specific chapters on these factors. The nutritional factors of importance are: the nature of the nutrients, especially the carbon and nitrogen sources; the concentrations of the substrates, particularly whether or not the substrate is growth-limiting; the supply of product precursors and of stimulants.

16.7.2 Carbon and energy sources

An excess of certain carbon and energy sources inhibits product formation perhaps by the mechanism known as *catabolite repression* (Demain, 1972b). Examples are inhibition by excess glucose of penicillin production and of diphtheria toxin production (Righelato & van Hemert, 1969b). The repression may be overcome by empirical selection of the carbon sources, for instance lactose in the penicillin fermentation or maltose in diphtheria toxin production, or by continuous feeding of the carbon source so that it is not present in excess. Mixed carbon sources may give better yields than single carbon sources. For instance adding acetate to a glucose medium increases the production of glutamic acid by *Corynebacterium* species (Aida, 1972, p. xviii). Frequently mixtures of carbohydrate and fat are found to be better than carbohydrate alone as a carbon source for antibiotic fermentations.

16.7.3 Nitrogen sources

The commonly used nitrogen sources are ammonia, nitrate, amino acids, peptone and proteins. Excess of ammonia inhibits glutamic acid production by *Corynebacterium* species and for this reason the ammonia source is added intermittently (Kinoshita, 1972, p. 267). In contrast a high ammonium ion concentration favours L-proline production by *Brevibacterium* species (Okumura, 1972). The complex nitrogen sources, such as corn steep liquor, various seed proteins or fish meal, often stimulate secondary metabolite production but the mechanism of the effect is not understood.

16.7.4 Nutrient deficiencies

Product formation is often stimulated by a nutrient deficiency. Phosphate limitation stimulates some antibiotic fermentations (Demain, 1972b). Deficiencies of trace metals stimulate the production of citric acid by *Aspergillus niger* and riboflavin by yeasts. It is necessary to ensure that the effect of a substrate limitation is not masked by an inhibitory effect, such as catabolite repression, caused by excess of another substrate.

A biotin deficiency stimulates the overproduction of glutamate by *Corynebacterium* species. This effect is attributed to an increase in the permeability of the plasma membrane to glutamic acid (Kinoshita, 1972, p. 314). In the presence of excess biotin a high concentration of glutamic acid occurs intracellularly but the plasma membrane prevents the escape of the amino acid from the cell (Demain, 1972b). The excretion of glutamic acid from the bacteria grown in the presence of excess biotin can also be induced by including either penicillin or a surfactant in the medium. Also these effects appear to be the result of disruption of the surface layers of the bacteria, which affects membrane permeability (Kinoshita, 1972).

16.7.5 Product precursors

Addition of a product precursor often stimulates product formation. A striking example is the stimulation of penicillin production by the addition of phenylacetic acid which is the side chain precursor of penicillin G. The precursor increases the q_{pen} several fold and leads to the exclusive production of penicillin G rather than several other different penicillins. Other examples of product precursors which stimulate fermentations are discussed by Demain (1972b).

16.7.6 Stimulants

The term 'stimulant' is given to compounds, which are not product precursors, but are able to stimulate the formation of specific products. The use of such stimulants has been reviewed by Perlman (1973). An outstanding example is the use of methionine or its analogue, norleucine, to stimulate the production of cephalosporin C (Drew & Demain, 1973). Barbiturates direct the rifamycin fermentation to produce one particular derivative (Margalith & Pagani, 1961). The microbial oxidation of cholesterol to androstadiene-3,17-dione is stimulated by chelating agents (Perlman, 1973). Citric acid production by *Aspergillus* is stimulated by methanol or propanol (Moyer, 1953); perhaps this is an effect on membrane permeability. In this connection it should be noted that surfactants stimulate both the overproduction of riboflavin by yeast (Demain, 1972a), and production of some enzymes involved in primary attack on a substrate (Reese, 1972). Benzylthiocyanate stimulates the chlortetracycline fermentation (Hostalek *et al.*, 1969). Production of demethyl chlortetracycline is stimulated by adding sulphonamides or ethionine to the culture to inhibit the methylating enzyme system. Benzimidazole inhibits the growth of *Saccharomyces cerevisiae* and causes excretion of xanthine (Evans & Brown, 1973). In general, the action mechanisms of stimulants are unknown; some may act by inhibiting or activating specific enzyme reactions.

CHAPTER 17

Effects of chemical inhibition and activation of growth

17.1 Introduction

Most studies on metabolic inhibitors have been focussed on their qualitative effects and the molecular mechanisms involved. Here particular attention is given to the quantitative effects and the possible use of inhibitors to modify fermentation processes (see also Section 16.7).

An inhibitor of growth is expected to cause both phenotypic change and selection of genetic variants which increase the organism's resistance to the inhibitor, that is maintain the growth rate. The model of an autosynthetic system (Section 24.2.1) predicts that transiently, after addition of an inhibitor, the growth rate will oscillate. Such behaviour was shown in the response of *Klebsiella aerogenes* to barbitone (2 g/l), which caused damped oscillations in the growth rate for about six generations (Dean & Moss, 1971). The Hinshelwood model of autosynthesis (Section 24.4) also predicts that a growth inhibitor will modify the enzyme activities of the cell so as to maximize the growth rate. In keeping with this prediction, the activities of certain pentose phosphate pathway enzymes were increased 2·4-fold in *K. aerogenes* by the addition of barbitone. A simple theory of microbial response to growth inhibitors can be based on the kinetics of enzyme inhibition. In this approach we make the simplifying assumption that the biomass behaves as an enzyme reacting with the growth-limiting substrate. In *competitive inhibition* it is assumed that the inhibitor competes with the growth-limiting substrate for uptake by the biomass. In *non-competitive inhibition* the inhibitor is assumed to react with the biomass at some site other than that for uptake of the growth-limiting substrate without affecting the affinity for the substrate. Inhibition, which may combine the features of both competitive and non-competitive types, is not included in the models, however, the effect of nalidixic acid on magnesium-limited growth of *Klebsiella* seems to be of this type since μ_m was decreased and K_s was increased (Dean & Moss, 1970). Besides inhibitors added to the medium, we may have inhibitory products and inhibitory substrates. There seems little justification for more elaborate models until the simple ones are proved inadequate by experimental tests.

An activator of growth is defined as a substance which, though not essential for growth, can increase the maximum growth rate. The existence of such growth activators is not clearly established, but there is evidence that some substances fall into this category.

17.2 Competitive inhibition

17.2.1 In batch culture

By analogy with enzyme kinetics (Dixon & Webb, 1967, p. 318) we have for the reaction of the biomass, X, with the growth-limiting substrate, S,

$$X + S \rightleftharpoons XS \qquad\qquad 17.1$$

from which

$$K_s = [X][S]/[XS] \qquad\qquad 17.2$$

where the square brackets denote concentration and K_s is the dissociation constant. For the combination of the biomass with the inhibitor, I, we have

$$X + I \rightleftharpoons XI \qquad\qquad 17.3$$

where it is assumed that I combines with the same enzyme site as does the substrate. From the dissociation of XI we obtain $K_i = [X][I]/[XI]$. The product, P, is formed according to the reaction

$$XS \rightarrow P + X \qquad\qquad 17.4$$

and this is considered to be the rate-limiting process, that is the rate of substrate breakdown $V = k[XS]$ where k is a constant. The total biomass is given by $x = [X] + [XS] + [XI]$. The specific rate of substrate utilization is given by $V/x = q$. From the expressions for the concentrations of the reactants, as in enzyme kinetics (Dixon & Webb, 1967, p. 318), we deduce that the specific rate of substrate utilization is given by

$$q = q_m s/(s + \alpha K_s) \qquad\qquad 17.5$$

where q_m is the maximum rate of substrate utilization, which is obtained when the biomass is saturated with substrate and $\alpha = 1 + i/K_i$ where i is the inhibitor concentration. If Y is the growth yield we substitute $q = u/Y$ and $q_m = \mu_m/Y$ then Eqn 17.5 becomes

$$\mu = \mu_m s/(s + \alpha K_s) \qquad\qquad 17.6$$

Equation 17.6 shows that, with competitive inhibition, K_s is apparently increased by the factor, α, but μ_m is unaltered. The inhibitor may affect the affinity for the growth-limiting substrate by combining with the biomass at a site different from that binding the substrate. In such a case the effect will be to increase K_s but the expression for α will be different (compare Dixon & Webb, 1967, p. 320). In batch growth, according to the model, the inhibitor essentially extends the decelerating growth phase, as depicted in Fig. 17.1.

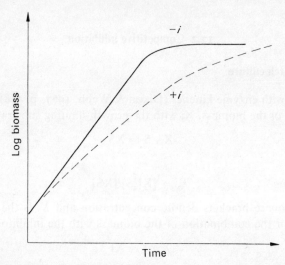

FIG. 17.1   Batch growth with $(+i)$ and without $(-i)$ competitive inhibitor added to medium.

17.2.2 In chemostat culture

If the competitive inhibitor is added to a chemostat culture with dilution rate, D, for the steady state we can substitute $\mu = D$ in Eqn 17.6 and we obtain

$$\tilde{s} = \alpha K_s D/(\mu_m - D) \qquad 17.7$$

$$\tilde{x} = Y(s_r - \tilde{s}) \qquad 17.8$$

where Y = growth yield and s_r = concentration of growth-limiting substrate in medium feed. The critical dilution rate is

$$D_{\text{crit}} = \mu_m s_r/(s_r + \alpha K_s) \qquad 17.9$$

The effect of the inhibitor, depicted in Fig. 17.2, is to lower D_{crit} and there is a more gradual decrease in the steady-state biomass level as D_{crit} is approached. A plot of $1/D$ versus $1/s$ with a given inhibitor concentration will have an intercept $-\alpha K_s$ on the $1/s$-axis. Rearrangement of Eqn 17.7 gives

$$\tilde{s} = c + ci/K_i \qquad 17.10$$

where $c = DK_s/(\mu_m - D)$, hence, if i is varied while D is held constant, a plot of $\tilde{s}$ against i will be a straight line with slope c/K_i and intercept c on the s-axis.

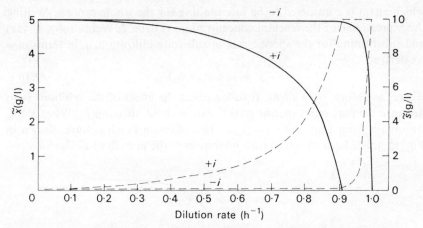

FIG. 17.2   Effects of competitive inhibitor added to culture medium of chemostat culture: $+i$, with inhibitor; $-i$, without inhibitor. *Parameters:* $s_r = 10$ g/l; $K_s =$
0.01 g/l; $i = 1$ g/l; $K_i = 0.01$ g/l; growth yield, $Y_x = 0.5$; $\mu_m = 1.0$ h^{-1}. Continuous
lines, steady-state biomass concentration, $\tilde{x}$; broken lines, steady-state concentration of growth-limiting substrate, $\tilde{s}$.

Van Uden (1967) verified that Eqn *17.10* holds for sorbose inhibition of
glucose uptake by a yeast under anaerobic conditions. The value of K_s for
glucose was 5.6×10^{-4} M, and the value of K_i, 1.8×10^{-1} M.

17.3 Non-competitive inhibition

17.3.1 In batch culture

With a non-competitive inhibitor it is assumed that the inhibitor combines
with the biomass at some site other than the one which binds the growth-
limiting substrate and we have the reactions

$$X + S \rightleftharpoons XS \qquad\qquad 17.11$$
$$X + I \rightleftharpoons XI \qquad\qquad 17.12$$
$$XS + I \rightleftharpoons IXS \qquad\qquad 17.13$$
$$XI + S \rightleftharpoons IXS \qquad\qquad 17.14$$

Reactions *17.12* and *17.13* are supposed to have the same dissociation constant, K_i; also Reactions *17.11* and *17.14* are supposed to have the same dissociation constant, K_s. It is assumed that the complex IXS cannot break down
into products, so that product comes only from the reaction

$$XS \rightarrow X + P \qquad\qquad 17.15$$

which again is considered to be rate-limiting for the whole process. Writing the expressions for the reactant concentrations (Dixon & Webb, 1967, p. 322) and substituting for the specific rate of substrate utilization, q, in terms of μ we derive

$$\mu = \mu_m s/\alpha(s + K_s) \qquad 17.16$$

where, as before, $\alpha = 1 + i/K_i$. It follows that the effect of the inhibitor is to decrease the maximum specific growth rate without affecting K_s. When $s \gg K_s$ we obtain from Eqn *17.16*, $\mu = \mu_m/\alpha$. This effect in batch culture, shown in Fig. 17.3, has been observed with inhibition of the growth of *Escherichia coli*

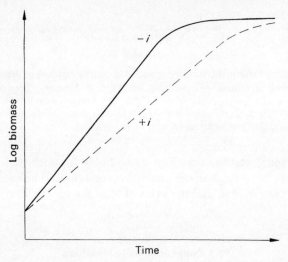

FIG. 17.3 Batch growth with $(+i)$ and without $(-i)$ non-competitive inhibitor added to medium.

by sulphonamides (Klotz & Gutman, 1945). The effect of the sulphonamide is antagonized by addition of the metabolic intermediate, *p*-aminobenzoic acid (PAB), and Klotz & Gutman showed that the specific growth rate (μ_i) with different concentrations of PAB and a given sulphonamide concentration, i, agreed with the expression

$$\mu_i = \mu_m b/(b + \alpha K_b) \qquad 17.17$$

where b is the concentration of PAB and α is given by $(1 + i/K_i)$. This result indicates that inhibition, which is non-competitive with respect to the growth-limiting substrate may be competitive with respect to an intermediary metabolite.

17.3.2 In chemostat culture

For the steady state in a chemostat culture with non-competitive inhibition we can substitute $\mu = D$ in Eqn 17.16 and obtain the following expressions for the steady-state concentrations of growth-limiting substrate and biomass:

$$\tilde{s} = DK_s \Big/ \left(\frac{\mu_m}{\alpha} - D\right) \qquad\qquad 17.18$$

$$\tilde{x} = Y(s_r - \tilde{s}) \qquad\qquad 17.19$$

and the critical dilution rate is given by

$$D_{\text{crit}} = \mu_m s_r / \alpha(s_r + K_s) \qquad\qquad 17.20$$

The effects of the inhibitor on the steady-state values are depicted in Fig. 17.4. When $s_r \gg K_s$ and $D_{\text{crit}} \approx \mu_m/\alpha$, then substituting $\alpha = 1 + i/K_i$ we obtain

$$i = K_i \frac{\mu_m}{D_{\text{crit}}} - K_i \qquad\qquad 17.21$$

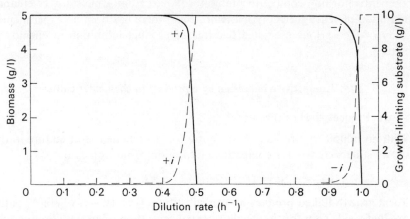

FIG. 17.4   Effect of non-competitive inhibitor added to culture medium of a chemostat culture: $+i$, with inhibitor; $-i$, without inhibitor. Continuous lines, steady-state biomass concentration; broken lines, steady-state growth-limiting substrate concentration. *Parameters:* $s_r = 10$ g/l; $K_s = 0.01$ g/l; $i = 0.01$ g/l; $K_i = 0.01$ g/l; $Y_x = 0.5$; $\mu_m = 1.0$ h^{-1}.

Hence, if the inhibitor concentration is varied, a plot of i against $1/D_{\text{crit}}$ should be a straight line with slope, $K_i \mu_m$ and intercept K_i. The parameter D_{crit} could be measured by the non-steady-state (wash-out) method.

Inhibition of growth of bacteria and yeasts by ethanol causes a decrease in μ_m in accord with non-competitive inhibition (Dean & Hinshelwood, 1966, p. 145; Aiba et al., 1968). The experiments of Aiba et al. (1968) with chemostat cultures of yeast show that ethanol does not affect the K_s value for glucose. Applying the model of Section **17.3.2** to the results of Aiba et al., the value obtained for the K_i of ethanol is 0.5 M.

17.4 Product inhibition

A fermentation product may inhibit growth by competing with the growth-limiting substrate for uptake, or the product may inhibit growth non-competitively. A product which may cause competitive inhibition of growth is ferric ion obtained in the oxidation of ferrous ion by *Ferrobacillus* (D. Kelly, private communication). Ethanol is an example of a product which causes non-competitive inhibition of the growth of yeast. Only the cases of inhibitory product formation where either the product yield or q_p is constant are treated here.

In batch culture, if the product inhibits growth either competitively or non-competitively, there will be a progressive decrease in the specific growth rate, and in the extreme case it may cause cessation of growth. In contrast, the theory predicts that steady-state chemostat cultures will show different features depending on whether or not the inhibitory product is growth-linked, or it is a competitive or non-competitive inhibitor. Apart from some study of ethanol inhibition of yeast growth (Aiba *et al.*, 1968) there has been no experimental study of the various predicted features of product inhibition in chemostat culture.

17.5 Competitive inhibition by a product in chemostat culture

17.5.1 Product yield (Y_p) constant

With an inhibitory product at concentration p in a chemostat culture in the steady state we have, for competitive inhibition, from Eqn *17.6*

$$\mu = D = \mu_m \tilde{s}/\{\tilde{s} + (1 + \tilde{p}/K_i)K_s\} \qquad 17.22$$

For a growth-linked product we can substitute $\tilde{p} = Y_p(s_r - \tilde{s})$ where Y_p is the product yield. Thus for the steady-state concentration of growth-limiting substrate we obtain

$$\tilde{s} = K_s D(K_i + Y_p s_r)/\{\mu_m K_i - D(K_i - K_s Y_p)\} \qquad 17.23$$

The steady-state biomass concentration is given by $\tilde{x} = Y_x(s_r - \tilde{s})$ where Y_x is the growth yield. The effect of the inhibitory product on the steady-state biomass is depicted in Fig. 17.5.

17.5.2 With constant q_p

If the inhibitory product is not growth-linked we express its steady-state concentration by $\tilde{p} = q_p \tilde{x}/D$ where q_p, the specific rate of product formation, is assumed to be constant. We also substitute $\tilde{x} = Y_x(s_r - \tilde{s})$, then from Eqn *17.22* we obtain

$$\tilde{s} = K_s(DK_i + q_p Y_x s_r)/\{K_i(\mu_m - D) + q_p Y_x K_s\} \qquad 17.24$$

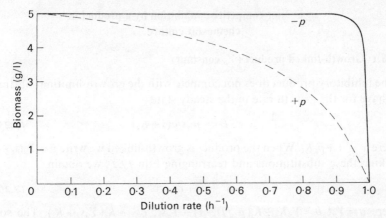

FIG. 17.5   Effect of competitive inhibition by product (Y_p constant) on steady-state biomass concentration in a chemostat culture: $+p$, with product; $-p$, without product. *Parameters:* $s_r = 10$ g/l; $K_s = 0.01$ g/l; $Y_x = 0.5$; $Y_p = 0.2$; $\mu_m = 1.0$ h^{-1}.

It is assumed that none of the growth-limiting substrate is used to form the inhibitory product. A plot of the biomass concentration against dilution rate is given in Fig. 17.6. Unlike the case of the growth-linked competitive in-

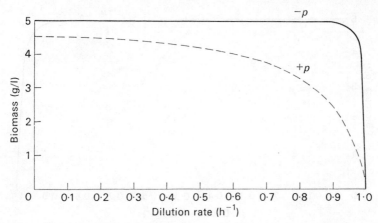

FIG. 17.6   Effect of competitive inhibition by product on steady-state biomass concentration in a chemostat culture when q_p is constant: $+p$, with product; $-p$, without product. *Parameters:* $s_r = 10$ g/l; $K_s = 0.01$ g/l; $Y_x = 0.5$; $q_p = 0.2$ g/g. biomass.h; $K_i = 0.01$ g/l; $\mu_m = 1.0$ h^{-1}.

hibitor as $D \to 0$ the value of $\tilde{s}$ approaches a finite value. However on decreasing the dilution rate the model must eventually fail since $\tilde{p}$ must approach a finite value limited by the concentration of some substrate.

17.6 Non-competitive inhibition by a product in chemostat culture

17.6.1 Growth-linked product (Y_p constant)

If the inhibitory product does not compete with the growth-limiting substrate, we have for the growth rate in the steady state

$$\mu = D = \mu_m \tilde{s}/\alpha(\tilde{s}+K_s) \qquad\qquad 17.25$$

where $\alpha = 1+\tilde{p}/K_i$. When the product is growth-linked we write $\tilde{p} = Y_p(s_r - \tilde{s})$. Making these substitutions and rearranging Eqn *17.25* we obtain

$$a\tilde{s}^2+b\tilde{s}+c = 0 \qquad\qquad 17.26$$

where $a = Y_p$; $b = Y_p K_s + K_i(\mu_m/D-1) - Y_p s_r$; $c = -K_s(Y_p s_r + K_i)$. The solution for $\tilde{s}$ is

$$\tilde{s} = -\frac{b \pm \sqrt{b^2-4ac}}{2a} \qquad\qquad 17.27$$

Of the two roots, one is negative, which in practice, can be neglected. The steady-state biomass is derived from $\tilde{x} = Y_x(s_r - \tilde{s})$. The predicted effect of the

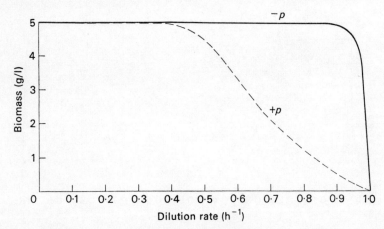

FIG. 17.7 Effect of non-competitive inhibition by a product (Y_p, constant) on steady-state biomass in a chemostat culture: $+p$, with product; $-p$, without product. *Parameters:* $s_r = 10$ g/l; $K_s = 0\cdot01$ g/l; $Y_x = 0\cdot5$; $Y_p = 0\cdot1$; $K_i = 1\cdot0$ g/l; $\mu_m = 1\cdot0$ h^{-1}.

inhibitory product on growth is shown (Fig. 17.7). The curve showing the biomass with inhibitory product is unusual in that it becomes convex towards the biomass axis at the higher dilution rates. For this convex part of the curve

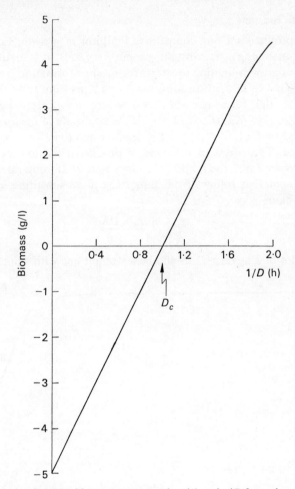

FIG. 17.8   Relation between biomass concentration $(\tilde{x})$ and $1/D$ for a chemostat culture with non-competitive inhibition by product $(Y_p,$ constant) when $\tilde{s} \gg K_s$; from the data given in Fig. 17.7. D_c = critical dilution rate.

with $\tilde{s} \gg K_s$, Eqn 17.25 approximates to

$$D = \mu_m/(1 + Y_p\tilde{x}/Y_xK_i) \qquad 17.28$$

which on rearrangement becomes

$$\tilde{x} = \mu_m Y_x K_i/Y_p D - Y_x K_i/Y_p \qquad 17.29$$

Thus a plot of $\tilde{x}$ against $1/D$ at high dilution rates should be a straight line with slope, $\mu_m Y_x K_i/Y_p$ and intercept, $-Y_x K_i/Y_p$ on the biomass axis; the intercept on the axis of $1/D$ will be $1/D_{\text{crit}}$. For the data of Fig. 17.7 the straight line relation of $\tilde{x}$ against $1/D$ holds from $D_{\text{crit}} = 1 \cdot 0$ to $D < 0 \cdot 6$ (Fig. 17.8).

17.6.2 With constant q_p

When the product is a non-competitive inhibitor of growth, Eqn *17.25* will apply and, assuming q_p is constant, we put $\tilde{p}=q_p\tilde{x}/D$. It is assumed that the utilization of growth-limiting substrate for product formation is negligible, so that $\tilde{x}=Y_x(s_r-\tilde{s})$. On substituting for $\tilde{p}$ and $\tilde{x}$ in Eqn *17.25*, it can be re-arranged in the form $a\tilde{s}^2+b\tilde{s}+c=0$ where $a=q_pY_x$; $b=K_i(\mu_m-D)-q_pY_x(s_r-K_s)$; $c=-K_s(DK_i+q_pY_xs_r)$. Of the two roots of the quadratic equation for $\tilde{s}$ (given by Eqn *17.27*) one is negative and therefore, in practice, can be neglected. The effects of this type of product inhibition are depicted in Fig. 17.9. When $\tilde{s}\gg K_s$ the graph of biomass against dilution rate is seen to be a straight line. This follows from Eqn *17.25* if we substitute $\alpha=1+q_p\tilde{x}/K_i$, when we obtain

$$\tilde{x} = \frac{\mu_m K_i}{q_p} - \frac{DK_i}{q_p} \qquad\qquad 17.30$$

Hence a plot of x against D should be a straight line with slope $-K_i/q_p$.

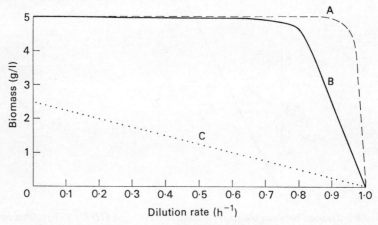

FIG. 17.9 Effect of non-competitive inhibitory product on steady-state biomass in a chemostat culture, when q_p is constant. Broken line A, without inhibitory product; *parameters:* $s_r=10$ g/l, $K_s=0\cdot01$ g/l, $Y_x=0\cdot5$, $\mu_m=1\cdot0$ h^{-1}. Continuous line B, with inhibitory product; parameters as for line A and with $q_p=0\cdot02$ g/g biomass h, $K_i=0\cdot5$ g/l. Dotted line C, with inhibitory product; parameters as for curve A and with $q_p=0\cdot04$ g/g. biomass. h, $K_i=0\cdot1$ g/l.

17.7 Inhibitor affecting growth yield

An inhibitor may alter the direction of metabolism so that a different growth yield is obtained in the presence of the inhibitor. In batch culture, when $s\gg K_s$, we have for the growth rate $\mu_m=q_mY_x$, where q_m is the maximum rate of uptake of growth-limiting substrate and Y_x is the growth yield. Suppose that Y_x is decreased to Y_x^i in the presence of inhibitor and there is no change in q_m, then the maximum growth rate will be decreased to $\mu_m^i = q_m Y_x^i$. In a chemostat culture the steady-state growth rate with inhibitor present is given by

$$D = \mu_m^i s/(s + K_s) \qquad\qquad 17.31$$

and the steady-state values of growth-limiting substrate and biomass concentrations are $\tilde{s} = K_s D/(\mu_m^i - D)$ and $\tilde{x} = Y_x^i(s_r - s)$. The effects of this type of inhibition in batch culture and chemostat culture are depicted in Fig. 17.10. This type of behaviour may be expected if uncouplers of oxidative phosphorylation are added to processes limited by supply of energy source.

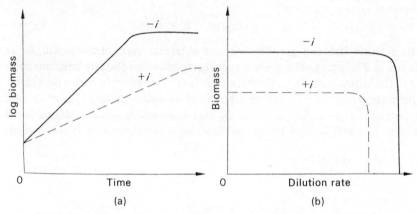

FIG. 17.10   Effects of an inhibitor which decreases the growth yield: (a) in batch culture; (b) in chemostat culture; $+i$, with inhibitor; $-i$, without inhibitor. The maintenance ration is assumed to be zero.

17.8 Substrate inhibition of growth

17.8.1 Effect on growth rate

Some substrates, for example alcohols, phenols and hydrocarbons, if present in excess also inhibit growth. A model of the effect can be based on the kinetics of substrate inhibition of enzyme activity (Dixon & Webb, 1967, p. 75). If X and S represent biomass and substrate we have the reaction

$$X + S \rightleftharpoons XS \qquad\qquad 17.32$$

for which $K_s = [X][S]/[XS]$. Inhibition of substrate utilization is supposed to be caused by the reaction of a second substrate molecule with the biomass substrate complex; this inhibitory reaction is represented by

$$XS + S \rightleftharpoons XS_2 \qquad\qquad 17.33$$

for which

$$K_i = [XS][S]/[XS_2] \qquad 17.34$$

Product is supposed to be formed by the breakdown of the complex XS according to the reaction

$$XS \to X + P \qquad 17.35$$

The rate of substrate utilization is supposed to be limited by the rate of the reaction given by Eqn 17.35; this rate is given by

$$V = k[XS] \qquad 17.36$$

where k is a constant. From expressions for the reactant concentrations we obtain

$$\mu = \mu_m s K_i / (s K_i + K_s K_i + s^2) \qquad 17.37$$

The relation between growth rate and substrate concentration will be as shown in Fig. 17.11. The growth rate increases with substrate concentration up to the critical value s_{crit}, thereafter the inhibitory effect becomes dominant. The maximum value of μ can be determined by differentiating Eqn 17.37 with respect to s; when $d\mu/ds = 0$, μ is at the maximum and $s_{\mathrm{crit}} = (K_s K_i)^{\frac{1}{2}}$. In a batch culture, in which initially the growth-limiting substrate concentration exceeds

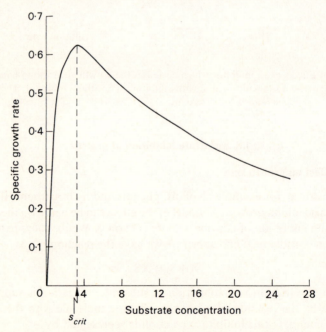

FIG. 17.11 Effect of inhibitory substrate on specific growth rate: $\mu_m = 1 \cdot 0$, $K_s = 1$, $K_i = 10$.

s_{crit}, the growth rate will increase until $s < s_{crit}$. Uemura *et al.* (1969) found that *n*-pentane was an inhibitory substrate for a bacterial species; the ratio, K_i/K_s was about 100. When K_i and K_s are so far apart, the maximum specific growth rate $\approx \mu_m$ and K_s and K_i approximate respectively to the lower and upper values of s when $\mu = \mu_m/2$.

17.8.2 Effect on chemostat culture

In a steady state of a chemostat culture the growth rate, μ of Eqn *17.37* can be equated with the dilution rate, D, thus we obtain an equation for $\tilde{s}$ of the form $a\tilde{s}^2 + b\tilde{s} + c = 0$ where $a = 1$; $b = (1 - \mu_m/D)K_i$; $c = K_iK_s$. The steady-state biomass concentration is given by $\tilde{x} = Y_x(s_r - \tilde{s})$ where Y_x is the growth yield and s_r is the concentration of growth-limiting substrate in the feed medium. The shape of the curves for $\tilde{x}$ and $\tilde{s}$ are shown in Fig. 17.12. Two steady states are

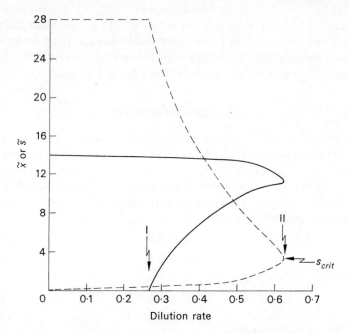

FIG. 17.12   Effect of inhibitory growth-limiting substrate in chemostat culture: $\mu_m = 1\cdot0$, $K_s = 1$, $K_i = 10$, $s_r = 28$, $Y_x = 0\cdot5$. Continuous line, steady state biomass concentration, $\tilde{x}$; broken line, steady-state concentration of growth-limiting substrate, $\tilde{s}$. Points I and II are the lower and upper critical dilution rates respectively.

possible between dilution rates I and II in Fig. 17.12. Of the two steady states, if $\tilde{s} < s_{crit}$, the growth rate will respond to an increase in s by increasing the growth rate, and vice versa, that is the culture is self regulating and the steady state is stable. But if $s > s_{crit}$, the steady state is not stable, because an increase in s will decrease the growth rate so that the culture will wash out. Also, when $s > s_{crit}$, if a disturbance decreases s, the growth rate will increase and the biomass will increase until it comes to the stable steady-state value. Stable steady states with $s > s_{crit}$ may be obtained in the second stage of two chemostats in series. The D_{crit} may vary over the dilution rate range from point I to point II in Fig. 17.12. At the lower value of D_{crit}, $\tilde{s} = s_r$ and at the higher value $\tilde{s} = s_{crit}$. When $\tilde{s} \gg K_s$ it follows from Eqn 17.37 that

$$\mathrm{I}/\mu = \mathrm{I}/\mu_m + \tilde{s}/\mu_m K_i \qquad\qquad 17.38$$

then a plot of I/μ or I/D against $\tilde{s}$ should be a straight line with slope, $\mathrm{I}/\mu_m K_i$, and intercept, I/μ_m.

Jones *et al.* (1973) found by application of two chemostats in series that growth of a bacterial species limited by phenol agreed with this model for substrate-limited growth. In this system, the K_s value for phenol was 0·9 mg/l and $K_i = 110$ mg/ml. Also it was found that with $\tilde{s} > s_{crit}$ the growth yield from phenol was lower (0·59) than it was when $\tilde{s} < s_{crit}$ ($Y = 0·9$). Since the maintenance energy (expressed as phenol consumed) was the same under either condition it follows that the variation in the overall growth yield, Y, was caused by variation in the true growth yield (Y_{EG} in Eqn 8.9).

17.9 Activators of growth

An activator of growth is defined as a substance which is not essential for growth and is not metabolized, but can intervene in the growth process so as to increase the maximum growth rate. The action of such a substance is modelled like that of an enzyme activator (Dixon & Webb, 1967, p. 323). If X, S, A and P represent the biomass, growth-limiting substrate, activator and product respectively we have

$$X + S \rightleftharpoons XS \qquad\qquad 17.39$$
$$X + A \rightleftharpoons XA \qquad\qquad 17.40$$
$$XA + S \rightleftharpoons XAS \qquad\qquad 17.41$$
$$XS \rightarrow X + P \qquad\qquad 17.42$$
$$XAS \rightarrow XA + P \qquad\qquad 17.43$$

From Reaction 17.40 we have

$$K_a = [X][A]/[XA] \qquad\qquad 17.44$$

where the square brackets denote concentrations. The overall reaction rate (V) is supposed to be limited by the breakdown of the complexes XS and XAS to form products, that is

$$V = k[XS] + k'[XAS] = qx \qquad\qquad 17.45$$

where q is the specific rate of substrate utilization and x is the total biomass present. When there is no activator present and the biomass is saturated with substrate then $[XS]=x$ and we have, for the maximum specific rate of substrate utilization, $q_m=k$. When the biomass is saturated with activator and substrate we put $[XAS]=x$, then the maximum specific rate of substrate utilization is given by $k'=q'_m$. Putting $q_m=\mu_m/Y_x$ and $q'_m=\mu'_m/Y_x$ where Y_x is the growth yield it follows that $k'/k=\mu'_m/\mu_m$. Substituting for $[XS]$ and $[XAS]$ in terms of x, s, K_s and K_a we obtain

$$\mu = \beta\mu_m s/(s+K_s) \qquad\qquad 17.46$$

where

$$\beta = \left(1+\frac{a}{K_a}\cdot\frac{\mu'_m}{\mu_m}\right)\bigg/\left(1+\frac{a}{K_a}\right)$$

Thus the apparent effect of the activator is to increase μ_m without affecting the K_s value.

The response of the human HeLa cell to insulin (Blaker *et al.*, 1971) might fit this model, since it was noted that insulin deficiency was characterized by a very low growth rate. Under magnesium limitation a *Bacillus* species secretes a substance which acts like an activator (Meers & Tempest, 1968). These authors suggested that the effect could be modelled by an expression similar to Eqn *17.46*. Examples of enzyme activators are chloride ions for alpha-amylase and certain divalent or trivalent anions for fumarate hydratase (Dixon & Webb, 1967, p. 443).

CHAPTER 18

Cultures at low and zero growth rates

18.1 Stationary phase behaviour

On cessation of growth, a culture enters into the stationary phase. The true stationary phase is caused by exhaustion of a nutrient, chemical inhibition of growth or a physical stress. The decline phase, characterized by autolysis and decrease in the biomass, may commence after a stationary phase or immediately after growth ceases.

The growing vegetative cell appears to be unstable when, for any reason, growth stops. Cessation of growth seems to initiate immediately a sequence of changes in the structure and function of the organism. Autolysis in the decline phase is an extreme manifestation of the instability of the organism after growth ceases. The behaviour of stationary-phase organisms probably does not follow one pattern, but rather a variety of patterns depending on the nature of the growth-limiting substrate or the inhibitory condition.

Since, in theory, the growth rate of a culture can, by means of the chemostat, be held at any value above zero, there arises the question, how near to zero does the specific growth rate have to be before the culture shows the characteristic stationary-phase behaviour? Or, does stationary-phase behaviour begin at a finite growth rate? This question is considered in Section 18.4.

18.2 Stationary phase of a bacterial culture

18.2.1 General properties

In the early stationary phase bacterial cell size reaches a minimum. In the late stationary or decline phase distorted or swollen cells called 'involution forms' often occur. This feature, perhaps, reflects damage by lytic enzymes either to the cell wall or plasma membrane, or poor regulation of residual synthesis of cell components. Loss of the ability to retain the Gram staining character is commonly shown by Gram-positive bacteria in the stationary phase. Also the resistance of bacteria to many forms of physical and chemical stress caused by, for instance, a hypotonic medium, shear, sudden cooling and high temperature, is greater in the stationary phase than in the logarithmic growth phase. These differences indicate that the structures of the organism in the two phases differ. In some of the bacterial groups the formation of *endospores* or *exospores* is characteristic of the stationary phase. Even in non-spore-forming types there is cytological evidence of the development of *microcysts* (Bisset, 1950) which could be a resistant, dormant form of the bacteria.

18.2.2 Metabolism

A feature of non-growing organisms is their loss of enzyme activities (Thurston, 1972). These enzyme losses may be caused by the turnover of protein which suddenly rises from a level of 0·5% or less in growing bacteria, to 5% per hour when growth ceases (Mandelstam, 1960); however, some enzymes seem to disappear at rates greater than the protein turnover rate (Thurston, 1972).

The endogenous metabolism of non-growing bacteria is characteristic of the type of bacteria (Dawes & Ribbons, 1964). When starved, aerobic suspensions of Gram-positive bacteria simultaneously oxidize their pools of peptides and amino acids along with stored carbohydrate. In contrast, *Escherichia coli* oxidizes first any stored glycogen and subsequently peptide material, although the total amount of free amino acids remains constant.

18.2.3 Viability

When bacteria are suspended in a culture medium, but starved of a nutrient essential to growth, they progressively die. If starved of a carbon and energy source, addition of the carbon and energy source accelerates death of the bacteria, a phenomenon termed 'substrate-accelerated death' (Postgate & Hunter, 1964; Strange & Dark, 1965). This effect is antagonized by mM cyclic AMP (Calcott & Postgate, 1972).

The labelling of biomass by ^{14}C has shown that addition of carbon substrates to buffered suspensions of algae, moulds and yeasts stimulates the degradation of endogenous substrates (Moses & Syrett, 1955; Miles & Pirt, 1969). This activity may contribute to substrate-accelerated death.

Maintenance of the viability of streptococci in non-growing aqueous suspensions is enhanced by magnesium ions (Thomas & Batt, 1968). Also the presence of an energy storage compound can prolong the cell viability (Strange *et al.*, 1961).

18.3 Stationary phase of a fungal culture

18.3.1 Without carbon and energy source

Trinci & Righelato (1970) induced the decline phase or autolysis in *Penicillium chrysogenum* by stopping the glucose feed to a chemostat culture in the steady state. Autolysis, which began without lag, was characterized by decrease in mycelial dry weight at a constant rate of 8% per hour. Protein and RNA decreased initially at a rate of about 5% per hour but after 6 hours the rates of loss decreased. Total carbohydrate did not vary significantly for 48 hours but thereafter decreased by about 0·5% per hour. The DNA decreased dramatically by some 75% in 12 hours and then remained roughly constant. There was evidence that nucleic acid bases accumulated in the medium. The hyphae largely emptied of cytoplasm leaving the hyphal walls and a collection of membranes, presumably from organelles. A few normal hyphal compartments remained present after 5 days. No conidia were formed as was the case when there was a slow feed of glucose (Section **18.3.2**). In a glucose-limited batch culture autolysis began if the exponential growth phase was ended by sudden exhaustion of the energy source.

18.3.2 With maintenance energy supplied

The effects of restricting the glucose feed to the maintenance ration in a chemostat culture of *Penicillium chrysogenum* were studied by Righelato *et al.* (1968). The culture growth was glucose-limited. During the first 12 hours, protein, RNA and DNA were extensively degraded. Subsequently the protein content of the mycelium remained fairly constant in amount whereas RNA and DNA were resynthesized and increased to amounts exceeding the initial values. The total carbohydrate increased slightly. The respiratory quotient decreased from 0·97 to 0·72 which suggests that a major change occurred in the oxidative metabolism. From 24 hours the submerged hyphae formed conidia at a rapid, constant rate, which was maintained for at least 120 hours.

In a chemostat culture of *Aspergillus nidulans*, when the growth was stopped by restricting the glucose feed, the changes in the culture were similar to those observed with *P. chrysogenum* except that conidia were not formed (Bainbridge *et al.*, 1971). It is concluded that cessation of growth in moulds by limitation of carbon and energy source causes a profound reorganization of the mycelium which may, in some circumstances, result in formation of conidia.

18.4 Minimum growth rate

A number of reports based on chemostat experiments suggest that steady growth rates cannot be maintained below a finite minimum value (Pirt, 1972b). Tempest *et al.* (1967) systematically studied the effects of decreasing, to an extremely low value, the dilution rate in a glycerol-limited culture of *Klebsiella aerogenes*. Their results show that the specific growth rate tended to a minimum value of 0·009 h^{-1} ($t_d = 80$ h) compared with a maximum specific growth rate of about 1·0 h^{-1} ($t_d = 0·69$ h). In an ammonia-limited culture the growth rate tended to a minimum of 0·007 h^{-1}. The fact that at dilution rates of less than 0·2 h^{-1} a large proportion of the bacteria were non-viable was taken into account in the calculation of the specific growth rate (Section 7.2.2). The validity of this correction may be questioned, since there could have been present in the chemostat culture dormant, that is non-growing, bacteria, which appeared viable when plated out. Hence the number of growing bacteria could have been overestimated and the minimum specific growth rate underestimated. At dilution rates below 0·06 h^{-1} both the RNA and DNA contents of the biomass decreased more sharply, with dilution rate and the maintenance energy apparently decreased (Pirt, 1972b). These observations show that some properties of the bacteria changed sharply at a specific growth rate of about 0·06 h^{-1}.

Koch & Coffman (1970) found that in a chemostat culture of *Escherichia coli* with a dilution rate of 0·029 h^{-1} the population appeared to consist of two fractions distinguished by their ability to induce β-galactosidase. In about two-thirds of the population the induction was completed in the normal 10 minutes, but in the other one-third it took about 3 hours to complete the induction. In contrast with a dilution rate of 0·09 h^{-1} the population showed no such heterogeneity and the enzyme was induced in 10 minutes.

To account for these observations on the changes in the properties of *Klebsiella* and *Escherichia* at low growth rates below about 0·06 h^{-1} (6% of μ_m) it is postulated that a part of the population differentiates into dormant cells, which are characterized by suspension of growth, minimum RNA and DNA contents (about 6% and 2% of the dry biomass respectively), minimal or zero maintenance energy, and lag before induction of β-galactosidase. It follows that, in order to determine the true specific growth rate at a low growth rate, it would be necessary to estimate the 'dormant' proportion of the population, not simply the proportion viable. The postulated differentiation into dormant cells could allow a minimum growth rate to be maintained by the growing fraction of the population.

18.5 Kinetics of sporulation in *Bacillus*

Formation of endospores by *Bacillus* is commonly associated with the stationary phase of a batch culture, however, by means of chemostat culture it has been shown that sporulation is negatively correlated with growth rate (Dawes & Thornley, 1970). The model of Dawes & Thornley (1970) to account for the kinetics of sporulation is discussed below.

It is supposed that there occur in the culture the following processes: (*i*) vegetative cell multiplication, (*ii*) initiation of spore formation, (*iii*) maturation of spores, and (*iv*) germination of mature spores to regenerate vegetative cells. It is assumed that once sporulation is initiated the organism must complete the sequence of events leading to maturation of a spore. If we represent the vegetative cell by X, the initiated spore by Y, and the mature spore by Z then the four processes are represented respectively by

$$(i)\ \mathrm{X} \to 2\mathrm{X},\ (ii)\ \mathrm{X} \to \mathrm{Y},\ (iii)\ \mathrm{Y} \to \mathrm{Z},\ (iv)\ \mathrm{Z} \to \mathrm{X}$$

It is further assumed that the processes (i), (ii) and (iv) are first order reactions with rate constants, μ, K and α respectively. Let n, y and ζ be the numbers per unit volume of vegetative cells, initiated spores and mature spores respectively, and let t_m be the time from spore initiation to spore maturation.

In a batch culture, in the infinitely small time interval dt, the increase in vegetative cell number per unit volume is given by

net increase = growth − spores initiated + spores germinated

that is

$$dn = \mu n.dt - Kn.dt + \alpha\zeta.dt \qquad\qquad 18.1$$

hence

$$dn/dt = (\mu - K)n + \alpha\zeta \qquad\qquad 18.2$$

If the germination rate, $\alpha\zeta$ is small compared with $(\mu - K)n$ then we have

$$dn/dt \approx (\mu - K)n \qquad\qquad 18.3$$

so that the growth will appear exponential but with the apparent specific growth rate $(\mu - K)$.

In a chemostat culture the balance for vegetative cell increase is given by

net growth = growth − initiation − washout + germination
rate   rate   rate   rate   rate

that is

$$dn/dt = \mu n - Kn - Dn + \alpha\zeta \qquad\qquad 18.4$$

hence

$$dn/dt = (\mu - K - D)n + \alpha\zeta \qquad\qquad 18.5$$

The balance for initiated spores is

net initiation rate = initiation rate − maturation rate − washout rate

that is

$$dy/dt = Kn - Kn_{(t-t_m)}\,e^{-Dt_m} - Dy \qquad\qquad 18.6$$

The term $n_{(t-t_m)}$ is the number of vegetative cells present at time $(t-t_m)$. During the interval t_m some of the initiated spores will be washed out and the fraction of the spores initiated at time $(t-t_m)$, which remain and become mature spores, is e^{-Dt_m} (Section 5.4). The balance for mature spores is

net increase rate = maturation rate − washout rate − germination rate

that is

$$d\zeta/dt = Kn_{(t-t_m)}\,e^{-Dt_m} - D\zeta - \alpha\zeta \qquad\qquad 18.7$$

In the steady state when $dn/dt = dy/dt = d\zeta/dt = 0$, $n_{t-t_m} = \tilde{n}$ then Eqn 18.5, 18.6 and 18.7 become

$$(\mu - K - D)\tilde{n} + \alpha\tilde{\zeta} = 0 \qquad\qquad 18.8$$

$$K\tilde{n}(1-\epsilon) - D\tilde{y} = 0 \qquad\qquad 18.9$$

$$K\tilde{n}\epsilon - (D+\alpha)\tilde{\zeta} = 0 \qquad\qquad 18.10$$

where $\epsilon = e^{-Dt_m}$. It is supposed also that we can measure the proportion of the total population (N) which consists of mature spores, that is in the steady state

$$\tilde{\theta} = \tilde{\zeta}/\tilde{N} = \tilde{\zeta}/(\tilde{n} + \tilde{y} + \tilde{\zeta}) \qquad\qquad 18.11$$

From Eqn 18.8, 18.9, 18.10 and 18.11 we derive

$$K = \frac{D\tilde{\theta}(D+\alpha)}{D(\epsilon - \tilde{\theta}) - \alpha\tilde{\theta}(1-\epsilon)} \qquad\qquad 18.12$$

$$\mu = D + K\{1 - \alpha\epsilon/(D+\alpha)\} \qquad\qquad 18.13$$

$$\tilde{n} = \tilde{N}/\left\{\frac{K}{D}(1-\epsilon) + 1 + \frac{1}{\alpha}(K+D-\mu)\right\} \qquad\qquad 18.14$$

$$\tilde{\zeta} = K\tilde{n}t/(D+\alpha) \qquad\qquad 18.15$$

$$\tilde{y} = \tilde{N} - \tilde{n} - \tilde{\zeta} \qquad\qquad 18.16$$

It also follows from the model that the frequency of spore production is Kn and the frequency of vegetative cell formation is μn, and, since these are the only two events which a vegetative cell can undergo, the probability of spore initiation is given by

$$\phi = Kn/(\mu n + Kn) = K/(\mu + K) \qquad\qquad 18.17$$

Dawes & Thornley (1970) tested the theory on sporulation of *Bacillus subtilis* in a chemostat. The germination rate (α) was determined on isolated spore samples and found to be about 0.05 h^{-1} at $37°$C. The value of the spore maturation time, t_m was determined by a novel application of chemostat culture. Following a step down in the dilution rate the spore initiation rate will undergo a step increase and, therefore, at a time t_m later the spore number will show a step increase (Fig. 18.1). Thus the maturation time of the *Bacillus* spores at $37°$ was found to be 4 hours. The spore initiation rate K was negatively correlated with the growth rate according to the equation, $K = 0.091 - 0.143\mu$. The proportion of spores and the probability of spore initiation in any particular bacterial cell are shown in Fig. 18.2. The experimental data are in good agreement with the model.

Dawes *et al.* (1969) also showed that from the distribution of the spore

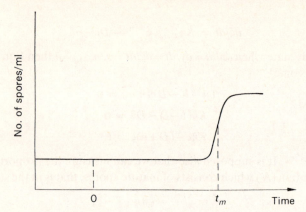

FIG. 18.1 Chemostat method of determining spore maturation time, t_m (Dawes & Thornley, 1970). At time zero the dilution rate is decreased so as to step up the spore initiation rate. After time, t_m there is a step up in the number of spores.

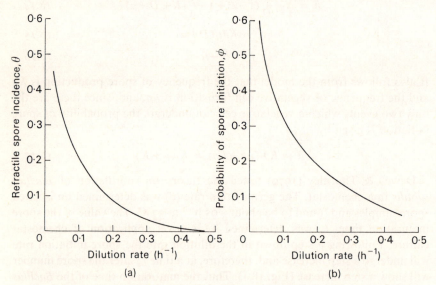

FIG. 18.2 (a) Proportion (θ) of mature spores and (b) probability (ϕ) of spore initiation in chemostat culture of *Bacillus subtilis*, glucose-limited at 37°C. (Redrawn from Dawes & Thornley, 1970)

forms in the different stages of development in a chemostat culture the duration of each phase of development is given by

$$\tau_{i-1} = t_m - \frac{1}{D} \ln \left(\frac{y_i}{y_r}\right) \qquad\qquad 18.18$$

where y_i = fraction of cells in stage X_i; y_r = fraction of cells showing refractile spores and τ_{i-1} = time after initiation to beginning of stage X_i. This study of bacterial sporulation is important as an example of how a transient process in a chemostat culture may quantitatively be analysed.

CHAPTER 19

Growth lag

19.1 Introduction

The nature and causes of lag in growth of bacterial cultures after inoculation attracted much attention during the 1930's, however, elucidation of the basic causes was held up until the mechanism of enzyme synthesis and its control had been discovered. Knowledge of the causes of lag are important because of the need to minimize or control its length. Alternatively the lag may be used as a parameter of the effects of environment on the organism.

Growth lag is a transition period during which the specific growth rate increases to the maximum value characteristic of the culture environment. The quantitative definition of growth lag is discussed in Section 2.8.

A 'true' lag is defined as a lag in the growth of the whole population. An 'apparent' lag results if a part of the population grows, from the start, at the maximum rate, and the other part fails to grow. During the lag there may be fluctuations in the specific growth rate. One type of fluctuation is a stepwise increase in cell numbers due to some synchronization of division. But normally such synchronization disappears after one or two generations. Subculture of starved bacteria or yeasts may cause a decrease in the viable biomass, which is a manifestation of substrate accelerated death (Section 18.2.3).

19.2 Apparent lag

Growth of only a part of the inoculum may be caused by death of some of the organisms or by a selective medium. If only a fraction α of the biomass is capable of growth then the growth rate of the biomass, x, is given by

$$dx/dt = \mu\alpha x \qquad\qquad 19.1$$

where $\mu\alpha$ is the 'apparent' specific growth rate. As x increases, through the addition of new, viable cells, $\alpha \to 1$ and $\mu\alpha \to \mu$, the true specific growth rate.

The apparent lag may indicate that only a few mutants or genetic variants in the population can grow. The long duration of such a lag, which may appear to be up to 20 generation times in a bacterial culture, can be used as a criterion that growth is due to selection of a few variants. An example is selection of

194

Escherichia coli able to grow on lactose in the presence of 3-fluoroglucose (Miles & Pirt, 1973).

The growth lag of mammalian cells in defined protein-free medium has been found to be much reduced by the presence of methylcellulose (Birch & Pirt, 1970). The mechanism of this effect is unknown.

19.3 Causes of true lag

The basic causes of true lag in growth are: (*i*) change in nutrition, (*ii*) change in physical environment, (*iii*) presence of an inhibitor, (*iv*) spore germination, (*v*) state of the inoculum culture. The first three factors may require some phenotypic change in the biomass to optimize it for the new condition, or the metabolism of the biomass may overcome the adverse environment, for example by a pH change.

A change in a nutrient will probably involve the induction of one or more new enzymes and the time required to synthesize the optimum amount of a new enzyme may be from 10 minutes up to many hours. In some species of organism the induction of enzymes to utilize a new carbon and energy source may not occur unless there is present a small amount of the original carbon and energy source. For example, the transition from glucose to lactose utilization in *Penicillium chrysogenum* will occur in mixtures of glucose and lactose but not if glucose-starved mycelium is added to a medium containing lactose as the sole carbon source. Adaptation of animal cells to utilize glutamic acid requires the presence of some glutamine (Blaker *et al.*, 1971).

The lag may be an expression of the time required to inactivate an inhibitor present in the medium. The toxic effects of trace metals will probably be more marked in a simple minimal medium than in a rich natural medium which contains substances such as amino acids that can complex metal ions. The toxicity of organic acids is often markedly pH dependent and decreases with increase in pH value. A substrate may be the inhibitor and the lag period will depend on the time taken to decrease the substrate concentration to the value which permits maximum growth rate (Section **17.8**). A product of the inoculum may inhibit growth. For instance, some members of the *Lactobacillaceae*, when subcultured from anaerobic into an aerobic medium may accumulate hydrogen peroxide, which is a potent growth inhibitor. The peroxide induces formation of peroxidase which removes the hydrogen peroxide. The lag caused by the peroxide may increase with inoculum size because the production of hydrogen peroxide is proportional to the cell concentration (Seeley & Vandemark, 1951).

Where the inoculum consists of spore forms the lag before vegetative growth represents the germination time of the spores.

19.4 Inoculum effects

Knowledge of the quantitative effects of inoculum 'age' on the growth lag in bacterial cultures is largely based on the study of *Klebsiella aerogenes* cultures by Lodge & Hinshelwood (1943). They arbitrarily measured the age of the inoculum from roughly the beginning of the logarithmic phase when the bacterial population was $10^6/\text{ml}$. The age effects, shown in Fig. 19.1, depended

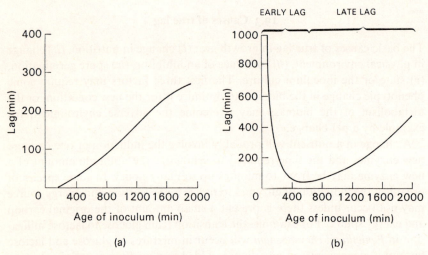

FIG. 19.1 Lag as a function of 'inoculum age' in cultures of *Klebsiella aerogenes*: (a) in glucose–asparagine–salts medium; (b) in glucose–ammonia–salts medium. (Redrawn from Lodge & Hinshelwood, 1943)

on whether the medium was minimal with glucose and ammonia as the sole carbon and nitrogen sources, or whether the ammonia was replaced by asparagine. In the glucose-asparagine medium a growth lag began to appear roughly when the inoculum source culture entered the stationary phase, and with increasing inoculum age the lag asymptotically approached a maximum. However, with increasing age, the lag became much less reproducible and may have been affected by death of the cells. The lag in the glucose–ammonia medium (Fig. 19.1b) was long when the inoculum source was at the beginning of its logarithmic growth phase. The lag reached a minimum when the inoculum source approached the end of the logarithmic phase, then increased with entry of the inoculum source culture into the stationary phase. Thus Lodge & Hinshelwood distinguished between the early lag which occurred with inoculum taken from the early logarithmic phase and late lag characteristic of inoculum from a stationary phase culture.

The early lag could be abolished either by the addition of sterile culture filtrate taken from a culture in the late logarithmic phase, or by increase in the inoculum size. In contrast, the late lag, or the lag in asparagine medium, was not affected either by the addition of culture filtrate or by increase in inoculum size. The fact that the early lag can be abolished by the addition of substances such as asparagine derived from citric acid cycle intermediates, or by increase in the inoculum size, is reminiscent of the means found to dispel the effects of a suboptimal dissolved carbon dioxide concentration (Section **8.10**). Air was bubbled through the *K. aerogenes* cultures used by Lodge & Hinshelwood so it seems possible that the cause of early lag was carbon dioxide deficiency. The causes of lag in the asparagine medium or of late lag in the minimal medium are discussed in Section **19.6**.

19.5 Effect of temperature

The effect of temperature on the growth lag in cultures of a *Staphylococcus* species in a rich medium is shown in Table 19.1. The nearly constant ratio of the growth lag to the doubling time suggests that the lag can be a useful parameter of the effects of environment on specific growth rate.

TABLE 19.1   Effect of temperature on growth lag of a *Staphylococcus* species in peptone medium (Cooper, 1963)

Temperature (°C)	Doubling time (t_d, min)	Lag (L, min)	(L/t_d)
25	64	120	1·9
30	38	66	1·7
35	29	50	1·7
40	27	39	1·5

19.6 Metabolic processes during lag

The lag associated with inoculum from the stationary phase of a culture is attributed to the reorganization necessary in the cell to reverse the changes caused by cessation of growth (Section **18.2**). The enzyme and RNA composition of the cell must be partly renewed during the lag. The model of Aiba *et al.* (1967) to account for growth lag in a yeast culture suggests that RNA synthesis determines the duration of the lag period. Thus the growth lag is an extreme case of the 'shift up' phenomenon described by Maaloe & Kjeldgard (1966, p. 98). These authors showed that a shift up in growth rate of bacteria must be preceded by a shift up in the RNA content of the organism.

Griffiths & Pirt (1967) found that in cultured mouse cells the pattern of amino acids uptake during the lag phase was much different from that during the logarithmic growth phase. In particular the specific uptakes of several amino acids were several times greater during the lag than during the logarithmic phase.

19.7 Conclusion

There are many causes of lag in growth. In order to avoid or minimize the lag the inoculum should be taken from a culture near to the end of the logarithmic phase in conditions as close as possible to those of the subculture medium. The size of the inoculum usually should be as large as is convenient but there is an exception to this rule mentioned in Section **19.3**. The addition of culture filtrate sometimes referred to as 'conditioned' medium may serve the same purpose as increasing inoculum size.

No studies on growth lag have taken account of the nature of the growth-limiting substrate in the inoculum. It seems probable that in most studies the carbon and energy source has been the growth-limiting substrate. Some of the effects associated with lag in batch culture may be simulated by a 'shift up' in growth rate in a chemostat culture.

CHAPTER 20

Mixed cultures

20.1 Introduction

The behaviour of mixed cultures, that is mixtures of different types of organisms, is important in the ecology of microbes in soil, water, disease and spoilage. Also, mixed cultures have important applications in the preparation of fermented foods and in the manufacture of microbial products, for example aspartic acid (Hotta & Takao, 1973). From the principles of mixed culture growth we can predict the outcome of contamination of a culture or the selection of a mutant type. Because of the complexity of mixed culture behaviour, mathematical models of the various systems are particularly valuable to describe and predict the behaviour.

This chapter is mostly a discussion of the possible patterns of behaviour of two species in a homogeneous culture. In a heterogeneous culture, such as a colony on a solid medium (Chapter 23) more complex effects may occur.

The final outcome of a mixed culture in a chemostat can be radically different from that in a batch culture. In a batch culture each species will be able to increase its biomass at a rate which will be a function of the chemical and physical environment, unless one species produces an agent which stops growth of the other, or there is a prey–predator interaction. In contrast, in a chemostat culture, all species which have a specific growth rate less than the dilution rate will decrease in amount and may therefore be eliminated from the culture.

The basic conditions for maintenance of two microbial species in a chemostat culture are defined in Table 20.1. A common condition is that the dilution rate must be less than the lower of the two critical dilution rates for each species. For mixed cultures in chemostats only the steady-state conditions are considered here; however, oscillations about the steady-state values do occur and are predicted by the models of some systems.

Various terms such as commensalism, symbiosis and parasitism are sometimes used to denote various types of interaction of species, however, there is much overlap in the meanings of these terms so that they do not fit any of the basic types of interaction defined in Table 20.1.

TABLE 20.1 Basic conditions for the maintenance of two microbial species in a chemostat culture

---

I   With same growth-limiting substrate
    a.   When specific growth rates coincide
    b.   When the faster growing species is inhibited by its own product
    c.   When a product of the faster growing species activates growth of the other species

II With different growth-limiting substrates
    a.   When the different growth-limiting substrates are fed into the culture
    b.   When a product of one species is the growth-limiting substrate for the other
    c.   When there is a prey–predator relationship

Note: In any system the dilution rate must be less than the critical dilution rate for both species.

20.2 Competition for the same growth-limiting substrate

20.2.1 Free competition

The outcome of competition between two species for the same growth-limiting substrate is determined by the influence of growth-limiting substrate concentration (s) on the specific growth rate (μ) of each species. If the relations of μ to s in each case were identical, no competition could occur, but probably the relations will differ in one of the ways shown in Fig. 20.1. In a batch cul-

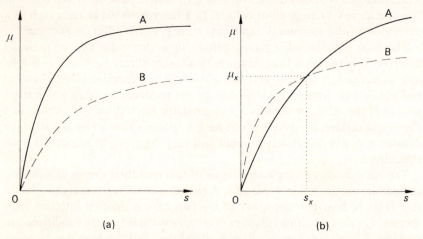

          (a)   (b)

FIG. 20.1 The possible relations between the specific growth rate (μ) and growth-limiting substrate concentration (s) of two species A and B: (a) without crossover, (b) with crossover.

ture, in which the initial concentration of growth-limiting substrate is large compared with the K_s value, each species will grow at its maximum rate until the limiting substrate is virtually exhausted. In contrast, in a chemostat culture, after the initial transient stage, the organism with the fastest growth rate would displace the other from the culture. If there is a 'crossover' in the relations of μ to s, as shown in Fig. 20.1b, then, in a chemostat, the faster growing species depends on the dilution rate. At the crossover point where $\mu = \mu_x = D_x$ and $\tilde{s} = s_x$, the two species could be maintained in a chemostat. Equating the expressions for μ (Eqn 2.21) we obtain

$$s_x = (\mu_{m(b)}K_a - \mu_{m(a)}K_b)/(\mu_{m(a)} - \mu_{m(b)}) \qquad 20.1$$

where $\mu_{m(a)}$ and $\mu_{m(b)}$ are the respective maximum growth rates, and K_a and K_b the corresponding saturation constants of species A and B.

The first instance of a 'crossover point' for two species in a chemostat culture was reported by Meers (1971) who found that in a magnesium-limited chemostat culture of *Bacillus subtilis* with *Candida utilis* at dilution rates below about $0.08\ \text{h}^{-1}$ the *Candida* species displaced the *Bacillus* species, whereas at higher dilution rates the *Bacillus* displaced the *Candida*. However, there was also an inoculum size effect in that, if the *Bacillus* was added at a concentration $< 10^8/\text{ml}$, the *Bacillus* was in any case eliminated. This inoculum size effect was explained by the need of the *Bacillus* to secrete in the medium some product which stimulated uptake of magnesium ions, and this product did not reach sufficient concentration at low bacterial density (Meers & Tempest, 1968).

20.2.2 Controlled competition

With class 1b conditions of Table 20.1, two species with the same growth-limiting substrate, are maintained in a chemostat if the faster growing species inhibits its own growth through a product (Fig. 20.2). Assuming that a product of species A competitively inhibits the uptake of the growth-limiting substrate by species A, and p is the product concentration, then the growth rate for the product-inhibited species is given (see Section 17.2) by

$$\mu_a = \mu_{m(a)}s/(s + \alpha K_a) \qquad 20.2$$

where $\alpha = (1 + p/K_i)$. For the other species we have

$$\mu_b = \mu_{m(b)}s/(s + K_b) \qquad 20.3$$

If $\mu_a > \mu_b$ when $p = 0$, then by increasing p it is possible for $\mu_a = \mu_b$, so that the two species may be maintained in a chemostat culture and the system should be selfregulating. Similar control would apply if the product were a noncompetitive inhibitor of the growth.

Under class 1c conditions of Table 20.1, two species competing for the same substrate are maintained in a chemostat culture if the faster growing species

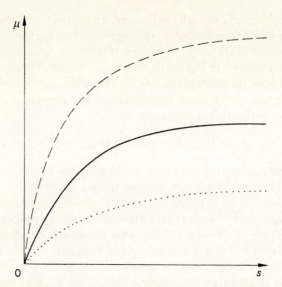

FIG. 20.2   Effects of an inhibitor or activator on the relations of specific growth rate
(μ) and growth-limiting substrate concentration (s) for a given species: continuous
line, without either inhibitor or activator; dotted line, with inhibitor; broken line,
with activator.

(A) produces a growth-activator for species B (see Fig. 20.2). For the $\mu - s$
relations we write for species A

$$\mu_a = \mu_{m(a)}s/(s + K_a) \qquad\qquad 20.4$$

and for species B

$$\mu_b = \beta\mu_{m(b)}s/(s + K_b) \qquad\qquad 20.5$$

where β is a function which increases with the concentration of the activatory
product (p), as discussed in Section 17.9. It is assumed that when $p = 0$, $\beta = 1$
and $\mu_b < \mu_a$; with increasing p, β increases until eventually $\mu_b = \mu_a$.

Brunner et al. (1968) reported that when Serratia marcescens and Escherichia
coli were apparently competing for the same growth-limiting substrate in a
chemostat, a stable mixed culture was obtained with E. coli dominant at the
lower dilution rates and S. marcescens dominant at the higher dilution rates.
The mechanism which permitted the two species to coexist in this case has not
been established.

20.3   Two species with different growth-limiting substrates

The common characteristic of the class II conditions for maintenance of two species in a chemostat culture is that there are different growth-limiting substrates at concentrations s_a and s_b respectively for species A and species B. This does not exclude the possibility that each species can utilize simultaneously both limiting substrates. The net accumulation rate for the growth-limiting substrate of species A is given by

$$ds_a/dt = D(s_{r(a)} - s_a) - \mu_a x_a / Y_{a/a} - q_b x_b \qquad 20.6$$

where the subscripts a and b indicate the parameters of species A and B respectively. The term $Y_{a/a}$ denotes the growth yield of species A from its growth-limiting substrate. The parameter q_b is the metabolic quotient for the consumption of the growth-limiting substrate of species A by species B. If $x_a / Y_{a/a} \gg q_b x_b$ then Eqn 20.6 approximates to

$$ds_a/dt = D(s_{r(a)} - s_a) - \mu_a x_a / Y_{a/a} \qquad 20.7$$

which is identical with the expression for limiting substrate consumption in a pure culture. If a similar equation holds for growth of species B on its limiting substrate, then the growth of the two species will be the sum of their separate growths on the growth-limiting substrates.

Such a system has been found to occur with *Pseudomonas aeruginosa* and *Klebsiella aerogenes* growing on glucose and *p*-hydroxybenzoate at 40°C and pH 7·2 when glucose is the limiting substrate for the *Pseudomonas* and hydroxybenzoate is the limiting substrate for the *Klebsiella* (Trilli & Pirt, unpublished); however, at 37°C or below, the *Klebsiella* utilizes both substrates and excludes the *Pseudomonas*. Another interaction in this system is that the hydroxybenzoate behaves as an inhibitory substrate for the *Klebsiella* but not for the *Pseudomonas*. Consequently, if the concentration of hydroxybenzoate rises to the inhibitory level for *Klebsiella*, the *Pseudomonas* could utilize more hydroxybenzoate and decrease its concentration to a non-inhibitory level for *Klebsiella*.

20.4 Product of one species is substrate for the other

In the class IIb conditions of Table 20.1, the growth-limiting substrate for one species is a product of the other species. An example is the growth of propionibacteria on lactic acid derived from growth of streptococci on lactose in milk. Other examples can be found in Shindela *et al.*, 1965; Yeoh *et al.*, 1968; Chao & Reilly, 1972; and Hotta & Takao, 1973.

Suppose that in a chemostat culture of two species, A and B, the growth of the species A is limited by a substrate supplied at concentration s_r in the medium. The concentration of species A should be the same as it would be in a single-member culture. If the growth-limiting substrate for species B is a product of species A present at concentration p in the medium, the accumulation rate of this product will be

$$dp/dt = q_p x_a - Dp - \mu_b x_b / Y_{b/p} \qquad 20.8$$

where q_p is the specific rate of product formation. Assuming that $Y_{p/a} = dp/dx_a$ and $q_p = Y_{p/a}\mu_a$, for the steady state we can put $\mu_a = \mu_b = D$, also $\tilde{x}_a = Y_{a/s}(s_r - \tilde{s})$. Substituting for q_p and x_a in Eqn 20.8 we obtain for the steady state

$$\tilde{x}_b = Y_{b/p}\{Y_{p/a}Y_{a/s}(s_r - \tilde{s}) - \tilde{p}\} \qquad 20.9$$

The predicted effects of dilution rate on the biomasses of the two species in a chemostat culture are depicted in Fig. 20.3(a). The effects if q_p is independent of μ_a are shown in Fig. 20.3(b).

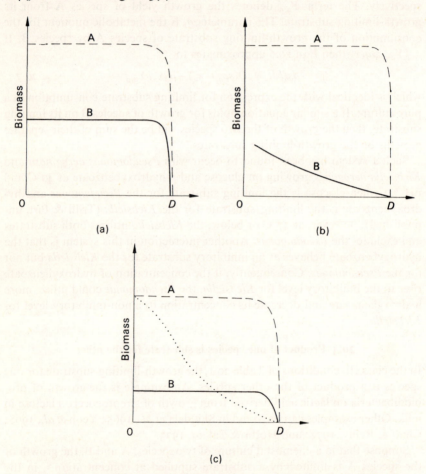

FIG. 20.3 Non-competitive growth of two species A and B as functions of the dilution rate (D) in a chemostat culture. (a) Species A forms a product which is the growth-limiting substrate for B ($Y_{p/a}$, constant). (b) Species A forms a product which is the growth-limiting substrate for B (q_p, constant). (c) Species A forms a product ($Y_{p/a}$, constant) which non-competitively inhibits its own growth and species B uses the inhibitor as growth-limiting substrate (dotted line shows growth of A in absence of B).

20.5 Inhibitory product of one species is the limiting substrate of the other species

It is supposed that species A produces an inhibitor of its own growth, which is present at concentration, p, and that species B utilizes the inhibitor as its growth-limiting substrate. For competitive inhibition by the product the growth rate of species A is given by

$$\mu_a = \mu_{m(a)}s/(s+\alpha K_s) \qquad 20.10$$

and for non-competitive inhibition

$$\mu_a = \mu_{m(a)}s/\alpha(s+K_s) \qquad 20.11$$

where $\alpha = (1+p/K_i)$. For the growth rate of species B we have $\mu_b = \mu_{m(b)}p/(p+K_b)$ then, in the steady state,

$$\tilde{p} = DK_b/(\mu_{m(b)}-D) \qquad 20.12$$

The steady-state value of s may be obtained from Eqn 20.10 or 20.11 putting $D=\mu_a$ and substituting for p from Eqn 20.12; $\tilde{x}_a = Y(s_r-\tilde{s})$ and $\tilde{x}_b$ is given by Eqn 20.9 for a growth-linked product. The values of the biomasses of species A and B as functions of dilution rate are depicted in Fig. 20.3(c).

An example of this type of mixed culture is the mixture of a methane-oxidizing *Pseudomonas* and *Hyphomicrobium* species which utilizes methanol (Wilkinson & Harrison, 1973). When oxidizing methane the pseudomonad accumulates methanol which is autoinhibitory and the *Hyphomicrobium* species utilizes the methanol thus overcoming the inhibition of the pseudo-monad.

20.6 Predator–prey interactions

The interaction of predators and prey is exemplified by protozoans ingesting bacteria. The system is modelled in the following way. Let us suppose that in a chemostat culture the predator is present at concentration, r. The prey, present at concentration, h, utilizes a growth-limiting substrate present at concentrations, s and s_r respectively in the culture and in the medium added. The relations between specific growth rate of prey (μ) and substrate concentration, and between the specific growth rate of predator (λ) and prey concentration are assumed to be of the hyperbolic form given by

$$\mu = \mu_m s/(s+K_s) \quad \text{and} \quad \lambda = \lambda_m h/(h+K_h) \qquad 20.13$$

where K_s and K_h are the saturation constants for prey and predator respectively. Evidence for the hyperbolic form of the $(\lambda-h)$ relation for growth of protozoa on bacterial prey has been given by Curds & Cockburn (1968). The growth yields are defined as Y g prey produced/g substrate consumed and W g predator/g prey consumed. The prey balance is

net rate of increase = growth rate − consumption rate − output rate
in prey

that is

$$dh/dt = \mu h - \lambda r/W - Dh \qquad 20.14$$

From the predator balance

net rate of increase in predator = growth rate − output rate

we obtain

$$dr/dt = (\lambda - D)r \qquad 20.15$$

The bacterial growth-limiting nutrient balance is

net rate of increase = input rate − consumption rate − output rate

that is

$$ds/dt = Ds_r - \mu h/Y - Ds \qquad 20.16$$

To obtain the steady-state values, the derivatives (dh/dt, dp/dt and ds/dt) are equated to zero. From Eqn 20.15 we find that in the steady state, $\lambda = D$, and from Eqn 20.14

$$\mu = D(1 + \tilde{r}/W\tilde{h}) \qquad 20.17$$

where the tilde denotes a steady-state value. Thus, in the steady state, the specific growth rate of the predator is adjusted to equal the dilution rate, and the specific growth rate of the prey is greater than the dilution rate as long as active predators are present. From Eqn 20.13 we obtain for the steady-state concentration of prey

$$\tilde{h} = DK_h/(\lambda_m - D) \qquad 20.18$$

By substituting for μ and $\tilde{h}$ in Eqn 20.16 when $ds/dt = 0$ we obtain

$$\tilde{s}^2 - \{s_r - K_s - \mu_m K_h/Y(\lambda_m - D)\}\tilde{s} - K_s s_r = 0 \qquad 20.19$$

hence $\tilde{s} = (-b \pm \sqrt{(b^2 + 4K_s s_r)})/2$ where $b = -\{s_r - K_s - \mu_m K_h/Y(\lambda_m - D)\}$. Also for the steady state we have the prey balance,

prey in effluent from = total prey produced − prey consumed by
culture   predator

that is

$$\tilde{h} = Y(s_r - \tilde{s}) - \tilde{r}/W \qquad 20.20$$

hence

$$\tilde{r} = W\{Y(s_r - \tilde{s}) - \tilde{h}\} \qquad 20.21$$

If the dilution rate exceeds the critical dilution rate, $D_{c(r)}$, for the predator, it will be displaced from the culture. At $D_{c(r)}$, the predator concentration $\tilde{r} = 0$

and $\tilde{h} = Y(s_r - \tilde{s})$, which approximates to Ys_r when $D_{c(r)} \ll \mu_m$ and $s_r \gg K_s$. Substituting $\tilde{h} = Ys_r$ in Eqn 20.13 we obtain

$$D_{c(r)} \approx \lambda_m Ys_r / (Ys_r + K_h) \qquad 20.22$$

If $Ys_r \gg K_h$, then $D_{c(r)} \approx \lambda_m$. Figure 20.4 shows the predicted steady-state values of p, h, s and μ as functions of dilution rate for a culture with bacteria as the prey and protozoa as the predators. Above the critical dilution rate for the protozoa the system behaves as a pure culture of the bacterial prey.

Canale (1970) and Curds (1971), from analysis and computer simulation of the differential equations for h, p and s, found that three possible states, which depend upon the growth parameters, may exist in the system: stable oscillations (limit cycles) or damped oscillations about the steady-state value, or an asymptotic approach to the steady-state value may occur. With growth parameters close to those used for Fig. 20.4, Curds (1971) found that stable oscillations were obtained when D was below about 75% λ_m, but with values of D nearer to λ_m the model system approached the steady-state value, at first with

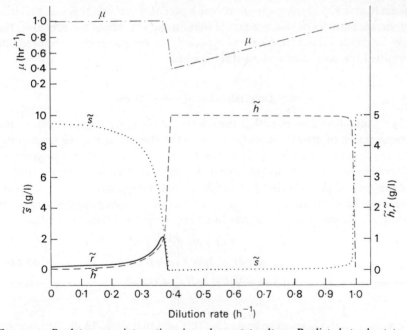

FIG. 20.4 Predator–prey interactions in a chemostat culture. Predicted steady-state values for concentrations of protozoan predator ($\tilde{r}$, g/l), bacterial prey ($\tilde{h}$, g/l) and growth-limiting substrate ($\tilde{s}$, g/l) for the bacteria. Specific growth rate of bacteria = μ; maximum specific growth rate of bacteria = 1.0 h^{-1}; maximum specific growth rate of protozoa = 0.4 h^{-1}; growth yield of protozoa = 0.5 g/g bacteria; growth yield of bacteria = 0.5 g/g substrate; $K_h = 0.10$ g/l; $K_s = 0.01$ g/l; $s_r = 10.0$ g/l. (From Pirt & Bazin, 1972)

damped oscillations, then asymptotically. Curds also found qualitative agreement between theory and experimental systems consisting of *Tetrahymena pyriformis* feeding on *Klebsiella aerogenes* for which sucrose was the growth-limiting substrate. In the predation of *Dictyostelium* species on *E. coli*, Tsuchiya *et al.* (1972) found that the populations of the two species oscillated over roughly a tenfold range so that on occasion the consumption of the growth-limiting substrate, which was glucose, was $< 10\%$ complete.

It is concluded that the predator can drastically reduce the prey population and consequently the extent of substrate utilization by the prey. This is of importance in effluent purification processes where, Pirt & Bazin (1972) suggested, protozoa might have an adverse effect by feeding on the bacteria which consume the waste material. However, the protozoa could be eliminated from the process by raising the dilution rate above the critical value for the protozoa. If necessary, the consumption of the bacteria by the protozoa could be arranged in a second-stage chemostat.

Jost *et al.* (1973) studied a culture consisting of two bacterial types, *Escherichia coli* and *Azotobacter vinelandii*, with glucose as growth-limiting substrate for each, and *Tetrahymena pyriformis* as a predator on both types of bacteria. Important features of this system are that, in the absence of the predator, the *E. coli* excluded the *Azotobacter* as expected, but the presence of the predator permitted the coexistence of all three species.

20.7 Determination of mutation rate

When a mutation does not affect the relation of the specific growth rate to the concentration of growth-limiting substrate the chemostat may be used to determine the mutation rate (Novick, 1958). Let N = total number of organisms/ml; m = number of mutants/ml (assumed small compared with N); μ = specific growth rate of parent type; ρ = specific growth rate of mutant; D = dilution rate; λ = rate of mutation expressed as number of mutants produced/organism produced. Hence in the chemostat in an infinitely small time interval dt

$$dm = \lambda\mu N\, dt + \rho m\, dt - Dm\, dt \qquad 20.23$$

$$\therefore \; dm/dt = \lambda\mu N + (\rho - D)m \qquad 20.24$$

$$dN = \mu N\, dt - \lambda\mu N\, dt - DN\, dt \qquad 20.25$$

$$\therefore \; dN/dt = \{(1 - \lambda)\mu - D\}N \qquad 20.26$$

We assume that λ is small so that $(1 - \lambda) \approx 1$ and that $\rho = \mu$. In the steady state we can substitute $\mu = \rho = D$ and from Eqn *20.24* obtain

$$dm/dt = \lambda DN \qquad 20.27$$

On integration Eqn *20.27* becomes

$$m = \lambda DNt + m_0 \qquad 20.28$$

Thus, on the assumptions made, the number of mutants rises linearly with time with a slope equal to λDN thus permitting λ to be estimated.

By this method Novick (1958) observed that, in chemostat cultures of *E. coli* B, there was a linear increase in the numbers of mutants resistant to phages T5 and T6. Also it was observed that, over a wide range, the rate of mutation was proportional to the generation time (t_d), that is $\lambda = \alpha t_d$ where α is the mutation rate per hour.

If μ is not equal to ρ, there will be selection for or against the mutant, and its number will not increase at a constant rate. Should $\rho > \mu$ the number of mutants will increase exponentially and eventually displace the parent organism. Should $\rho < \mu$, then, in the steady state, when $\mu = D$, the term $(\rho - D)$ in Eqn *20.24* is negative and as m increases $dm/dt \rightarrow 0$. Then the number of mutants will rise to a limiting value (m_L) obtained by putting $dm/dt = 0$ in Eqn *20.24*, that is

$$m_L = \lambda N/(1 - \rho/D) \qquad 20.29$$

20.8 Conclusion

There have been relatively few systematic tests of the principles of growth of mixed cultures. The interactions between species can best be tested by means of chemostat culture since many of the interactions will not be apparent unless the growth is substrate-limited and the specific growth rates of the organisms can be varied.

The six basic conditions which permit the coexistence of two species are listed in Table 20.1. Combinations of the basic conditions may help to stabilize the system. If the interactions between two species are mediated by a diffusible product, the system may be simplified by separating pure cultures of the two species by means of a dialysis membrane (Section **21.5**). By an extension of the class IIa condition of Table 20.1, it can be envisaged that any number (n) of species can coexist in a chemostat culture if, for each species, there is a different growth-limiting substrate, that is there are n different growth-limiting substrates. For this purpose a great variety of carbon, nitrogen or sulphur sources are available, but for the other essential elementary nutrients, principally phosphate, potassium and magnesium, there are few alternatives to the free ions. This leads to the conclusion that maximum diversity of population will occur where there is a multiplicity of carbon, nitrogen and sulphur compounds present with excess amounts of phosphate, potassium and magnesium ions. Conversely, minimum diversity should occur when either phosphate, potassium or magnesium are the growth-limiting substrates. Another way to maintain two competing species in the system is to have present a predator species.

The relation of the specific growth rate (μ) to the growth-limiting substrate concentration (s) is fundamental in determining the outcome of a mixed culture. In a chemostat culture, the higher the dilution rate, the fewer will be the number of species which can exist under homogeneous conditions. However, the ($\mu - s$) relation may be reversed and infinitely modified by the presence of inhibitors, activators and changes in temperature or other physical conditions.

CHAPTER 21

Batch cultures with substrate feeds

21.1 Fed batch culture

21.1.1 General description

The term 'fed batch culture' was introduced by Yoshida et al. (1973) to refer to a batch culture which is fed continuously with nutrient medium (Fig. 21.1). If a portion of the culture is withdrawn at intervals the system becomes a 'repeated fed batch culture', which can be maintained indefinitely. The volume variation in a fed batch culture distinguishes it from a chemostat culture in which it is essential to maintain the culture volume constant. Fed batch culture has been developed empirically for some industrial fermentations including production of penicillin, bakers' yeast and waste disposal by fermentation. Yoshida et al. (1973) found that fed batch culture increased the biomass yield from a hydrocarbon nearly twofold over that obtained by simple batch culture, an effect which remains unexplained. Restriction of the rate of substrate utilization by means of substrate feed rate is a means of overcoming 'catabolite repression' of product formation (Demain, 1972b). In bakers' yeast production the oxygen demand is controlled by means of the sugar feed rate. Fed batch culture also simulates some naturally occurring microbial systems such as the infected urinary tract (Mackintosh, 1973). The theory of fed batch culture (Pirt, 1974) shows that fed batch culture should have unique and important applications in the control of fermentations.

21.1.2 The quasi-steady state

Suppose that we have a homogeneous batch culture in which growth is limited by the concentration of one substrate and all other substrates are present in excess. Let s_r = initial concentration of growth-limiting substrate and x = biomass concentration at time, t. Then we have

$$x = x_0 + Y(s_r - s) \qquad 21.1$$

where x_0 = inoculum concentration, s = concentration of residual growth-limiting substrate and Y is the growth yield. It is assumed that when the bio-

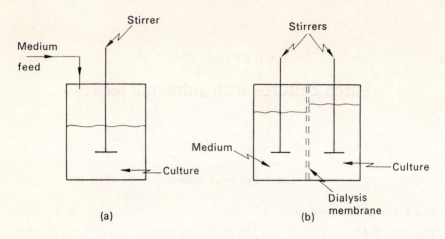

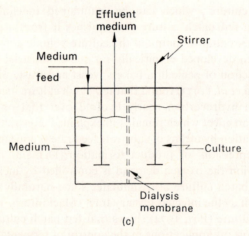

FIG. 21.1   (a) Fed batch culture. (b) Batch culture dialysed against a batch of medium.
    (c) Batch culture dialysed against a stream of medium.

mass concentration reaches its maximum value, x_m, the growth-limiting substrate is practically exhausted so that $s \ll s_r$. Then assuming the inoculum is small compared with the final biomass, we have $x_m \approx Y s_r$.

Now suppose that when $x_m \approx Y s_r$, a medium feed is started at flow rate F and with the growth-limiting substrate at concentration, s_r, in the feed medium. The total biomass in the culture is given by $X = xV$ where V is the culture volume at time, t. Since $x = X/V$, differentiation of this quotient gives for the growth rate

$$dx/dt = (V\,dx/dt - X\,dV/dt)/V^2 \qquad\qquad 21.2$$

We can substitute $dX/dt = \mu X$ where μ is the specific growth rate, $dV/dt = F$ and $F/V = D$ where D is the dilution rate. Then Eqn *21.2* becomes

$$dx/dt = (\mu - D)x \qquad\qquad 21.3$$

This equation is generally true for a fed batch culture. It is assumed that the relation of the specific growth rate to the concentration of growth-limiting substrate is of the Monod type, that is

$$\mu = \mu_m s/(s + K_s) \qquad\qquad 21.4$$

When $s_r \gg K_s$ over most of the range of μ from zero upwards, the growth-limiting substrate will be almost completely utilized so that when $x = x_m \approx Y s_r$, $dx/dt \approx 0$. With this condition it follows from Eqn *21.3* that $\mu \approx D$.

Let $S = $ total growth-limiting substrate in the culture, then for the substrate balance we have

rate of increase = rate of input − rate of consumption for growth

that is

$$dS/dt = F s_r - \mu X/Y \qquad\qquad 21.5$$

When $X = V x_m$ virtually all of the substrate is consumed as fast as it enters the culture so that $F s_r \approx \mu X/Y$, hence dS/dt and ds/dt are approximately zero. This state when $dx/dt \approx 0$, $ds/dt \approx 0$ and $\mu \approx D$ is called the *quasi-steady state*.

To obtain the concentration of growth-limiting substrate as a function of dilution rate in the quasi-steady state we substitute $D \approx \mu$ in Eqn *21.4* which gives

$$s \approx D K_s/(\mu_m - D) \qquad\qquad 21.6$$

The rate of increase in the total biomass during the quasi-steady state is given by

$$dX/dt = F Y s_r \qquad\qquad 21.7$$

hence

$$X = X_0 + F Y s_r t \qquad\qquad 21.8$$

Comparing a fed batch culture in the quasi-steady state with a chemostat culture in a steady state, in both cases in effect we have $\mu = D$ but, whereas D is constant in a chemostat, in a fed batch culture D is decreasing and μ is assumed to decrease at the same rate. The unique feature of fed batch culture is that, in the quasi-steady state, the biomass is in a transient state with the growth rate under control.

21.1.3 Effect of maintenance energy

We have assumed so far that the growth yield Y is constant. However, if the growth-limiting substrate is an energy source a part will be used to provide maintenance energy. Then for the energy balance we have

$$dS/dt = Fs_r - \mu X/Y_G - mX \qquad\qquad 21.9$$

where Y_G is the true growth yield and m is the maintenance coefficient. Provided the maintenance ration mX is small compared with the growth ration $(\mu X/Y_G)$ again we can put $Fs_r \approx \mu X/Y_G$, and consequently $\mu \approx D$. Putting $dS/dt = 0$ and substituting $D = \mu$ and $X = Ys_r$ where Y is the overall growth yield, from Eqn 21.9 we obtain

$$1/Y \approx 1/Y_G + m/D \qquad\qquad 21.10$$

Hence fed batch culture provides a means of estimating the maintenance energy and predicting the growth yield at different dilution rates.

When the growth-limiting substrate is the energy source, the total amount of biomass in the culture will have a maximum value given by

$$X_m = Fs_r/m \qquad\qquad 21.11$$

21.2 Product formation

21.2.1 With product yield ($Y_{p/x}$) constant

In the quasi-steady state of a fed batch culture with $x_m \approx Ys_r$ the product concentration will be $p \approx Y_{p/s}s_r$, where $Y_{p/s}$ is the product yield based on growth-limiting substrate consumed. The output rate of the product will be $F Y_{p/s}s_r$.

21.2.2 With constant q_p

In this case the rate of increase in the total amount (P) of a product in the quasi-steady state is given by

$$dP/dt = q_p x_m V = q_p x_m (V_0 + Ft) \qquad\qquad 21.12$$

where V_0 is the culture volume at time zero. On integration we obtain

$$P = P_0 + q_p x_m (V_0 + Ft/2) t \qquad\qquad 21.13$$

where P_0 is the total amount of product at time zero. Let p and p_0 be the product concentrations at times t and zero respectively, and substitute $pV = P$, $p_0 V_0 = P_0$, $D = F/V$, hence we obtain

$$p = p_0 V_0/V + q_p x_m (V_0/V + Dt/2) t \qquad\qquad 21.14$$

Sometimes the separate values of q_p and x_m may not be known because of difficulty in determining the biomass. However, if it is found that the rate of product synthesis is a constant, $r = q_p x_m$, we can make this substitution in Eqn 21.14 and obtain

$$p = p_0 V_0/V + r(V_0/V + Dt/2) t \qquad\qquad 21.15$$

This equation was applied by Pirt (1974) to estimate the possible penicillin titres in industrial fed batch cultures.

21.2.3 With q_p a complex function of μ

When the q_p value for a product varies in a complex way with the growth rate in a quasi-steady state of a fed batch culture, for the increase in total product in the small time interval dt, we write

$$dP = q_p(t)X\,dt \qquad\qquad 21.16$$

where $q_p(t)$ is a function giving the value of q_p at time, t. For X at time t in the quasi-steady state we substitute

$$X = x_m V = x_m(V_0 + Ft) \qquad\qquad 21.17$$

Then the total amount of product formed in time, t will be

$$P - P_0 = x_m \int_0^t q_p(t)(V_0 + Ft)\,dt \qquad\qquad 21.18$$

Substituting, $pV = P$ and $p_0 V_0 = P_0$ we obtain

$$p = \frac{p_0 V_0}{V} + \frac{x_m}{V}\int_0^t q_p(t)(V_0 + Ft)\,dt \qquad\qquad 21.19$$

If we substitute $r(t) = q_p(t)x_m$ in Eqn 21.19 we obtain

$$p = \frac{p_0 V_0}{V} + \frac{1}{V}\int_0^t r(t)(V_0 + Ft)\,dt \qquad\qquad 21.20$$

In order to evaluate the integral we need to determine the value of q_p as a function of t during a quasi-steady state. This may be done as follows. The product concentration, $p = P/V$. Differentiating this quotient we obtain

$$dp/dt = \frac{V(dP/dt) - P(dV/dt)}{V^2} \qquad\qquad 21.21$$

Substituting, $dP/dt = q_p X$, $X = Vx_m$, $P = p/V$, $dV/dt = F$ and $F/V = D$ we obtain

$$dp/dt = q_p x_m - Dp \qquad\qquad 21.22$$

hence

$$q_p = (dp/dt + Dp)/x_m \qquad\qquad 21.23$$

If we put $q_p x_m = r$, we obtain

$$r = dp/dt + Dp \qquad\qquad 21.24$$

To determine q_p or r experimentally in the quasi-steady state of a fed batch culture we plot p against t and from the slope of the graph obtain dp/dt at various times. Then, by means of Eqn 21.23 and 21.24, we can obtain graphical expressions of $q_p(t)$ and $r(t)$. Thus the integrals in Eqn 21.19 and 21.20 may be evaluated graphically.

21.3 Repeated fed batch culture

In a repeated fed batch culture part of the culture is removed at intervals. This means that the culture volume and consequently the dilution rate and dependent metabolic parameters such as specific growth rate will undergo cyclical variation. We will consider the case when the cycle time (t_w) is constant. It is assumed that during the process the culture is in a quasi-steady state. When the culture volume reaches a certain value, V_w a fixed fraction of the culture is removed so that the residual volume, $V_0 = \gamma V_w$. Substituting this value for V_0 in Eqn 21.14 we obtain for the product concentration at the time of culture withdrawal, that is the end of the cycle,

$$p_w = \gamma p_0 + q_p x_m \left(\gamma + \frac{D_w t_w}{2} \right) t_w \qquad\qquad 21.25$$

where $D_w = F/V_w$ is the dilution rate at the end of the cycle. The cycle time is the time for the culture volume to increase from γV_w to V_w, that is $t_w = (V_w - \gamma V_w)/F$. Since $F = D_w V_w$ the cycle time is given by

$$t_w = (1 - \gamma)/D_w \qquad\qquad 21.26$$

Substituting this value for t in Eqn 21.25 we obtain for the product concentration at the end of the cycle

$$p_w = \gamma p_0 + \frac{q_p x_m}{2 D_w} (1 - \gamma^2) \qquad\qquad 21.27$$

If q_p is not constant but varies throughout the cycle it follows from Eqn 21.19 that the product concentration at the end of the cycle is given by

$$p_w = \gamma p_0 + x_m \int_0^{t_w} q_p(t)(\gamma + D_w t)\, dt \qquad\qquad 21.28$$

If a repeated fed batch culture in a quasi-steady state has any product concentration, p_0, at the beginning of the first cycle, the product concentration after the first cycle is given by

$$p_0 = \gamma p_0 + K \qquad\qquad 21.29$$

where

$$K = q_p x_m (1 - \gamma^2)/2 D_w \qquad\qquad 21.30$$

when q_p is constant, or

$$K = x_m \int_0^{t_w} q_p(t)(\gamma + D_w t) \, dt \qquad 21.31$$

when q_p varies throughout the cycle. After the second cycle, the product concentration is

$$p_2 = \gamma p_1 + K = \gamma^2 p_0 + \gamma K + K \qquad 21.32$$

After the nth cycle, the product concentration is

$$p_n = \gamma p_{n-1} + K = \gamma^n p_0 + K(\gamma^{n-1} + \gamma^{n-2} + \ldots + \gamma + 1) \qquad 21.33$$

On summing the geometrical progression with constant ratio γ we obtain

$$p_n = \gamma^n p_0 + \frac{K(1 - \gamma^{n-1})}{1 - \gamma} \qquad 21.34$$

When n is large, $\gamma^n p_0 \to 0$ and $\gamma^{n-1} \to 0$, then Eqn 21.34 approximates to

$$p_n = K/(1 - \gamma) \qquad 21.35$$

Thus it is concluded that, irrespective of its initial value, the product concentration tends towards the value p_n (given by Eqn 21.35) in a repeated fed batch culture.

If q_p is constant so that we can substitute for K from Eqn 21.30 we obtain

$$p_n = q_p x_m (1 + \gamma)/2 D_w \qquad 21.36$$

Now, as we decrease the volume of culture removed at the end of the cycle, $\gamma \to 1$, and it follows from Eqn 21.36 that $p_n \to q_p x_m / D_w$, which is the product concentration expected in a chemostat steady state. Hence, if q_p is constant the product concentration obtained by repeated batch culture cannot exceed that obtained in the chemostat with $D = D_w$. However, the quasi-steady state conditions may stimulate the formation of some products so that the q_p exceeds the chemostat steady-state value (Section **21.4**).

21.4 Applications of fed batch culture

Fed batch culture is one means of achieving substrate-limited growth. By this means Yoshida *et al.* (1973) obtained hexadecane-limited growth of *Candida tropicalis* and found that the growth yield (0·95) obtained under the substrate-limited condition was almost twice that obtained in a simple batch culture; the cause of the difference in yield is unexplained. Bainbridge *et al.* (1971) used the method to study the changes in composition and structure of *Aspergillus nidulans* at zero growth rate, obtained by limiting the glucose feed to the maintenance ration. The ability to restrict the substrate feed rate is an advantage when the oxygen transfer rate (K_La) can limit the process rate. An important ecological application of the method has been to simulate a human bladder infection (Mackintosh, 1973). The observation by Pirt (1971) that feeding a growth-limiting substrate at a constant rate results in a linear increase in the total biomass is evidence in support of Eqn 21.8. A major tenet of the theory is that, in the quasi-steady state, the organism automatically adjusts its specific growth rate so that it remains equal to the dilution rate. This condition is most likely to break down at the beginning of the cycle when a sudden shift up in the specific growth rate is demanded. The results of Mateles *et al.* (1965) suggest that in a chemostat culture of bacteria the specific growth rate can adjust immediately to a shift up in μ of about 20% μ_m but takes about three doubling times to adjust to a shift up of about 40% μ_m. The mathematical model of autosynthesis (Section 24.2.1) suggests that the rate of adjustment of μ in a quasi-steady state of a fed batch culture would not be the same as that of a steady-state culture in a chemostat, because the enzymic constitutions in the two states will differ. More data on the effects of sudden changes in dilution rates is required. If necessary, the sudden shift up in dilution rate at the end of a cycle could be avoided by pumping out the culture at a controlled rate.

The ability to restrict the minimum specific growth rate may be necessary to prevent decay of the synthetic activity of the biomass, as for instance in penicillin production (Pirt & Righelato, 1967). Informal reports indicate that the penicillin fermentation is now run industrially as a repeated fed batch culture developed on a purely empirical basis. The minimum specific growth rate to prevent decay of the penicillin synthetic activity is $0.014\,h^{-1}$ (Pirt & Righelato, 1967). The possible penicillin titres in a repeated fed batch culture are discussed by Pirt (1974).

The most important feature of a fed batch culture is that it is a unique means of realizing transient conditions between fixed growth rates. There is evidence that the maximum rates of some processes can be achieved only transiently, examples are production of some bacterial antigens (Pirt *et al.*, 1961) and penicillin production (Wright & Calam, 1968).

Fed batch culture is technically simpler than chemostat culture because the need to keep the culture volume constant, which often is the most exacting part of chemostat culture, is eliminated.

21.5 Dialysis culture

21.5.1 Batch dialysis

In this culture system, depicted in Fig. 21.1(b), substrate from a batch of medium is allowed to diffuse across a dialysis membrane into the culture. The rate of diffusion of solute across the membrane is taken to be proportional to the concentration difference (Schultz & Gerhardt, 1969). In the system shown in Fig. 21.1(b) it is assumed that the medium is maintained homogeneous by some form of agitation or by circulating the medium and culture past the membrane.

Let $A=$ area of dialysis membrane, $s=$ concentration of growth-limiting substrate in culture, $s_m=$ concentration of growth-limiting substrate in medium outside culture, $s_{m(0)}=$ concentration of growth-limiting substrate in medium outside culture at time zero, $X=$ total biomass at time t, $X_0=$ total biomass when $t=0$, $V_c=$ volume of medium inside culture, $V_m=$ volume of medium outside culture. Suppose that at time zero the diffusion of substrate across the membrane becomes growth-limiting, then the biomass growth rate is given by

$$\frac{dX}{dt} = \psi A Y(s_m - s) \qquad\qquad 21.37$$

where $\psi=$ permeability coefficient of the membrane for the particular substrate and Y is the growth yield. It is assumed that when growth is limited by the diffusion rate of the substrate, $s \ll s_m$ so that we can put $(s_m - s) \approx s_m$, then

$$dX/dt = \psi A Y s_m \qquad\qquad 21.38$$

hence the specific growth rate is

$$\mu = \psi A Y s_m / X \qquad\qquad 21.39$$

From the substrate balance in the medium compartment, when the growth is diffusion limited,

$$V_m \, ds_m/dt = -\psi A s_m \qquad\qquad 21.40$$

Integrating Eqn 21.40 gives

$$s_m = s_{m(0)} \, e^{-\phi t} \qquad\qquad 21.41$$

where $\phi = \psi A / V_m$. Substituting for s_m in Eqn 21.38 and integrating gives

$$X = X_0 + V_m Y s_{m(0)}(1 - e^{-\phi t}) \qquad\qquad 21.42$$

If there is no maintenance requirement for the growth-limiting substrate then, as $t \to \infty$, the biomass approaches the maximum given by

$$X_m = X_0 + V_m Y s_{m(0)} \qquad\qquad 21.43$$

The total amount of a growth-linked product present at time t will be

$$P = Y_{p/x}X = Y_{p/x}X_0 + V_m Y_{p/s}s_{m(0)}(1 - e^{-\phi t}) \qquad\qquad 21.44$$

where $Y_{p/x}$ and $Y_{p/s}$ are the product yields based on biomass produced and substrate utilized respectively. When q_p is constant we have $dP/dt = q_pX$. Substituting for X and integrating we obtain

$$P = P_0 + q_p\left\{(X_0 + V_mYs_0)t + \frac{V_m^2Ys_0}{\psi A}(e^{-\phi t} - 1)\right\} \qquad 21.45$$

If the product cannot diffuse across the membrane, the final concentration of product will be

$$p = P/V_c \qquad 21.46$$

however if the product is diffusible then its final concentration will be

$$p = P/(V_c + V_m) \qquad 21.47$$

Hence dialysis culture concentrates non-diffusible products by the factor, $(V_c + V_m)/V_c$.

21.5.2 Dialysis against a stream of medium

In the system depicted in Fig. 21.1(c), the medium outside the culture is supposed to be completely mixed and fresh medium is added at rate F; s_r =concentration of growth-limiting substrate in medium feed. The rate of change in the concentration of growth-limiting substrate in the medium outside the culture will be

$$V_m\,ds_m/dt = Fs_r - Fs_m - \psi A(s_m - s) \qquad 21.48$$

When growth is diffusion-limited it is assumed that $s_m \gg s$. With this condition, Eqn 21.48 approximates to

$$V_m\,ds_m/dt = Fs_r - (F + \psi A)s_m \qquad 21.49$$

Integrating Eqn 21.49 gives

$$s_m = [Fs_r - \{Fs_r - (F + \psi A)s_{m(0)}\}\,e^{-\phi t}]/(F + \psi A) \qquad 21.50$$

where $\phi = (F + \psi A)/V_m$. As $t \to \infty$, $e^{-\phi t} \to 0$ then, from Eqn 21.50, it follows that

$$s_m \approx Fs_r/(F + \psi A) \qquad 21.51$$

The rate of increase in the total biomass when growth is diffusion-limited and $s \approx 0$ is given by

$$dX/dt = \psi As_m Y \qquad 21.52$$

With increase in time s_m tends to the value given by Eqn 21.51 and the rate of increase in total biomass approaches

$$(dX/dt)_\infty = \psi A Y Fs_r/(F + \psi A) \qquad 21.53$$

that is the total biomass tends to increase at a constant rate.

In this system the total biomass should increase until either the culture is so dense that mixing becomes impossible, or until the rate of diffusion of substrate across the membrane becomes equal to the maintenance ration. The concentration of a product may be derived in a manner similar to that for batch dialysis.

21.5.3 Applications of dialysis culture

Basically there are three applications for dialysis culture. It provides another means for achieving substrate-limited growth. It should be noted that in batch dialysis culture, although growth may be substrate-limited, the specific growth rate decreases with time, in contrast to a chemostat culture in the steady state. The second application is to concentrate biomass and non-diffusible products; examples are production of vaccines and enzymes. Gallup & Gerhardt (1963) showed that *Serratia marcescens* could be grown to a concentration of 9 g dry weight/100 ml in a dialysis culture as compared with 0·8 g dry weight/ 100 ml without dialysis. Thus the method provides a means of growing organisms to a high biomass density with dilute substrates. The third application is to lower the concentration of a diffusible product inhibitory to growth. For example, thiobacilli could be grown to a high density on glucose provided an inhibitory product, probably pyruvic acid, was removed from the culture by dialysis (Borichewski & Umbreit, 1966).

Biphasic culture is a term applied by Tyrell *et al.* (1958) to a culture system which consists of liquid medium overlaying a gelled medium. This system is similar to batch dialysis except that the resistance to diffusion of substrate out of the gel increases with time.

21.6 The diffusion capsule

The diffusion capsule (Fig. 21.2) provides a means of feeding substrate at a constant rate into a culture, especially on the shake-flask scale (Pirt, 1971). The capsule has a small orifice sealed with a dialysis membrane. When the capsule is filled with a solution of the substrate and immersed in an aqueous medium, the substrate diffuses out for a long period at a constant rate, which is proportional to the initial substrate concentration. This apparent contravention of Fick's law of diffusion is unexplained.

If the rate of diffusion of substrate from the capsule is G g/h, then for the rate of increase in the total biomass in the culture we have

$$X = X_0 + GYt \qquad 21.54$$

where Y is the growth yield. Let $V=$culture volume and $x=$biomass concentration, then substituting $xV = X$ in Eqn 21.54 we obtain

$$x = x_0 + GYt/V \qquad 21.55$$

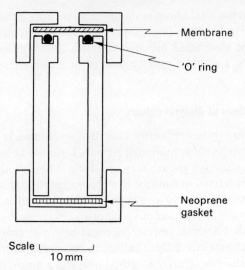

Membrane

'O' ring

Neoprene gasket

Scale └────────┘
        10 mm

FIG. 21.2   Diffusion capsule to feed substrate at a constant rate.

The validity of Eqn *21.55*, which indicates a linear increase in the biomass concentration, was established by Pirt (1971). The specific growth rate, it follows from Eqn *21.54*, is given by

$$\mu = GY/Vx \qquad\qquad 21.56$$

The growth-limiting substrate concentration can be obtained by substituting $\mu = \mu_m s/(s+K_s)$ in Eqn *21.56*, hence

$$s = GYK_s/(\mu_m Vx - GY) \qquad\qquad 21.57$$

and substituting for x by means of Eqn *21.55* we obtain

$$s = K_s/\{\mu_m(t+Vx_0/GY)-1\} \qquad\qquad 21.58$$

The product concentration at time t when q_p is constant will be

$$p = p_0 + q_p t(x_0 + GYt/2V) \qquad\qquad 21.59$$

O'Sullivan & Pirt (1973) used this expression to calculate the q_p for penicillin production in shake-flask cultures where growth was limited by glucose fed from a diffusion capsule.

CHAPTER 22

Submerged films and pellets of biomass

22.1 Submerged film of biomass

Many microorganisms and tissue cells in culture have the ability to stick to solid surfaces and multiply there. Thus bacteria and fungi in long-term stirred cultures will often coat the surfaces of the culture vessel. An outstanding example of the application of biomass films is the 'trickling filter' type of effluent purification. Mammalian and other types of animal tissue cells in static cultures attach to a glass surface, where they flatten and multiply in that form. Such a culture of animal cells is called a *monolayer culture*.

Initially the attached cells will each be fully exposed to the nutrient medium and be able to grow and multiply at the maximum exponential rate. However, when the cells form a thick confluent film of biomass, the growth will be limited by the diffusion of substrate into the film. A simplified treatment of diffusion-limited growth in a biomass film (Fig. 22.1) is given below.

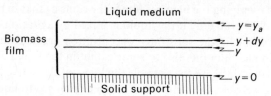

FIG. 22.1   Biomass film on solid surface. The concentration of limiting substrate at level y is s, at level $y+dy$, $s+ds$ and at level y_a, s_r.

We make the simplifying assumption that there is a steady state, that is

substrate entering the film = substrate metabolized by the film

For the lower region of the film from $y=0$ to y we have the balance

$$aD'\frac{ds}{dy} = qay\rho \qquad\qquad 22.1$$

where a=cross sectional area of film, D'=diffusion coefficient of substrate, q=metabolic quotient of growth-limiting substrate, ρ=density of biomass. Integrating Eqn 22.1 we obtain

$$D'\int_s^{s_r} ds = q\rho \int_y^{y_a} y\,dy \qquad\qquad 22.2$$

223

hence

$$D'(s_r - s) = \frac{q\rho}{2}(y_a^2 - y^2) \qquad\qquad 22.3$$

It is assumed that q is independent of s, a simplifying assumption which is valid for carbon and energy substrate concentrations above about 20 mg/l and for oxygen concentrations above 0·2 mg/l. Suppose $s=0$ when $y=0$, then for the maximum thickness of the active biomass layer we have

$$y_0 = (2D's_r/q\rho)^{1/2} \qquad\qquad 22.4$$

The maximum thickness of a respiring film of frog muscle, not oxygen-limited, with one side exposed to air at 1 atm pressure, is estimated to be 900 μm (Hill, 1928). In contrast, the maximum thickness of a film of *Escherichia coli*, not oxygen limited, when one side is exposed to air at 1 atm is estimated to be about 40 μm (Pirt, 1967). This value is much lower than that for the frog tissue mainly because the respiration rate is much greater. For fungal growth limited by glucose concentration (10 g/l) the thickness of the active layer is of the order of 1000 μm (Pirt, 1966). If the thickness of the biomass film exceeds the maximum thickness of the active film, it is assumed that below the 'active layer' either the layer is inactive or a different limiting substrate is being used. Thus the upper layer in contact with the liquid could be oxygen limited and the lower anaerobic layer, carbon limited.

In the model it is assumed that the concentration of limiting substrate in the liquid medium overlying the biomass film is the same as that in the bulk of the liquid. This will be true only if the liquid is mixed by agitation or convection currents. In still media the substrate concentration gradient may not be entirely in the biomass film but also to some extent in the liquid above the film. This would further decrease the possible thickness of the active biomass layer. If the process is aerobic, the diffusion of oxygen from gas to liquid will be important (Section 22.2.5).

22.2 Packed column with microbes attached

22.2.1 General features

Columns packed with solid materials to provide surfaces on which microbes can grow are found in trickling filters used for effluent purification, in the 'quick vinegar' process for acetification of alcohol, and in soil.

In the model proposed by Pirt (1973a), the column is envisaged as a series of 'theoretical films' of biomass (Fig. 22.2). Liquid medium flows down the column and provides nutrients to the biomass films. If the process is aerobic or evolves some gas the void between the packing material will be occupied partly by air or other gases.

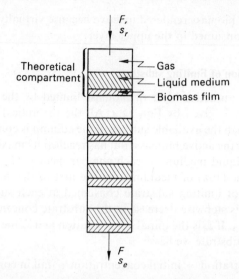

<figure>
Fig. 22.2   Diagrammatic representation of column of 'theoretical films' of biomass separating theoretical compartments each containing a volume V/n of liquid medium where n is the number of theoretical compartments and V is the total liquid volume in the column; F = medium flow rate, s_r = initial substrate concentration, s_e = substrate concentration in effluent from column.
</figure>

22.2.2 Assumptions in the theory

According to Pirt (1973a) the column is represented as n theoretical compartments each of which contains liquid medium overlying a basal theoretical film of biomass with cross sectional area (A) equal to that of the column. The liquid medium (volume, V) in the column is supposed to be distributed equally between the theoretical compartments. Substrates dissolved in the liquid will penetrate the biomass film only by diffusion. The liquid medium is supposed to percolate down the column at flow rate, F, so that it spends an equal time (t_r/n) in each compartment where t_r is the residence time in the entire column, that is there is no distribution of residence times. Thus the movement of liquid down the column is envisaged as 'intermittent' plug flow; it approaches plug flow more perfectly as the number of theoretical films increases. It is also assumed that in each compartment the liquid is completely mixed and there is no mixing of media from different compartments.

The biomass which originates by growth on the substrates in the medium will take some time to develop a film on a fresh column. With time the thickness of the film will increase, however, the thickness of the active layer of biomass will be limited by diffusion. Therefore, eventually, the biomass film will consist of an active upper layer in contact with the medium and below there may be a layer of biomass rendered inactive because virtually all of the limiting substrate is consumed in the upper layer.

22.2.3 Consumption of limiting substrate

The thickness of an active layer of biomass limited by the diffusion of an essential substrate is given by Eqn 22.4. All the theoretical films will come fully into play when the available surface in the column is completely covered with biomass and the active biomass film has reached its maximum thickness. The amount of liquid medium in each compartment is (V/n). Consider the passage of volume (V/n) of medium from the first to the last compartment. Let the amount of limiting substrate consumed at each successive step be $\sigma_1, \sigma_2, \ldots, \sigma_n$. This stepwise decrease in the substrate concentration is represented in Fig. 22.3. If s_r is the initial concentration and s_e the final concentration of limiting substrate we have

final concentration = initial concentration − fall in concentration

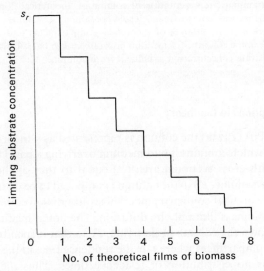

FIG. 22.3 Predicted mode of fall in concentration of growth-limiting substrate in a packed column with microbes attached.

that is

$$s_e = s_r - (\sigma_1 + \sigma_2 + \cdots + \sigma_n)/(V/n) \qquad 22.5$$

Now

$$\sigma_1 = q x_{gl} t_r / n \qquad 22.6$$

where q is the metabolic quotient for the growth-limiting substrate, x_{gl} is the amount of biomass in the active layer, t_r is the total residence time of medium in the column, t_r/n is the residence time in each compartment. The amount of biomass in the active layer will be

$$x_{gl} = Ay_1\rho \qquad\qquad 22.7$$

where A is the cross-sectional area of the column and y_1 is the thickness of the active layer; $\rho=$ biomass density. Substituting for y_1 from Eqn 22.4 and for x_{gl} from Eqn 22.7 we obtain

$$\sigma_1 = \frac{k}{n} s_r^{1/2} \qquad\qquad 22.8$$

where $k = At_r(2D'\rho q)^{1/2}$.

The concentration of limiting substrate in the second compartment is given by

$$s_r - \sigma_1/(V/n) = s_r - ks_r^{1/2}/V \qquad\qquad 22.9$$

The substrate consumed in the second theoretical compartment will be

$$\sigma_2 = \frac{k}{n}\left(s_r - \frac{k}{V}s_r^{1/2}\right)^{1/2} \qquad\qquad 22.10$$

that is

$$\sigma_2 = \frac{k}{n} s_r^{1/2}(1 - k/Vs_r^{1/2})^{1/2} \qquad\qquad 22.11$$

Putting $\zeta = k/Vs_r^{1/2}$, from the binomial theorem, provided ζ is a small fraction, $(1-\zeta)^{1/2} \approx 1 - \zeta/2$. Hence it is derived that σ_2 can be approximated to

$$\sigma_2 = \frac{k}{n} s_r^{1/2}\left(1 - \frac{k}{2Vs_r^{1/2}}\right) \qquad\qquad 22.12$$

Similarly it is derived that the substrate metabolized in the compartment of the nth theoretical film is

$$\sigma_n = \frac{k}{n} s_r^{1/2}\left\{1 - (n-1)\frac{k}{2Vs_r^{1/2}}\right\} \qquad\qquad 22.13$$

The sum of the series of σ is given by

$$\sum_{n=1}^{n} \sigma_n = \frac{k}{n} s_r^{1/2}\left[n - \frac{k}{2Vs_r^{1/2}}\{1+2+3+\ldots+(n-1)\}\right]$$

$$= ks_r^{1/2}\left\{1 - \frac{k(n-1)}{4Vs_r^{1/2}}\right\} \qquad\qquad 22.14$$

Since $\sum_{n=1}^{n} \sigma_n$ is the total limiting substrate consumed from a volume V/n of medium during its passage through the column we can put

$$\sum_{n=1}^{n} \sigma_n = (s_r - s_e)V/n \qquad\qquad 22.15$$

From Eqn 22.14 and 22.15 we obtain

$$\frac{k^2}{4V^2}n^2 - \left(\frac{k}{4V} + s_r^{1/2}\right)\frac{k}{V}n + s_r - s_e = 0 \qquad 22.16$$

Solving Eqn 22.16 gives

$$n = 0.5 + \frac{2F}{K}\left[s_r^{1/2} - \left\{\left(\frac{K}{4F} + s_r^{1/2}\right)^2 - s_r + s_e\right\}^{1/2}\right] \qquad 22.17$$

where $K = A(2D'\rho q)^{1/2}$. Equation 22.17 is called the *long formula*; if $K/4F$ is small compared with $s_r^{1/2}$ then Eqn 22.17 can be approximated to the *short formula*, that is

$$n = \frac{2F}{K}(s_r^{1/2} - s_e^{1/2}) = \frac{2V}{t_r K}(s_r^{1/2} - s_e^{1/2}) \qquad 22.18$$

The total biological film area is given by nA and the 'specific biological film area' is

$$\alpha_b = nA/V_c \qquad 22.19$$

where V_c is the volume of the packed column.

22.2.4 Biomass production

The total active biomass on the column is given by

$$x_g = F(s_r - s_e)/q \qquad 22.20$$

If we write $q = \mu/Y_g + \beta$ where μ/Y_g represents the rate of substrate utilization for growth only and β is the consumption for other purposes such as maintenance energy, then

$$x_g = \frac{F(s_r - s_e)}{(\mu/Y_g + \beta)} \qquad 22.21$$

The total biomass (x_t), that is active + inactive biomass, increases at the rate

$$dx_t/dt = YF(s_r - s_e) \qquad 22.22$$

where Y is the overall growth yield. Unless there is some process such as autolysis to remove the inactive biomass, it will tend to fill the void volume in the column. If the rate of loss of inactive biomass is represented by $b(x_t - x_g)$ where b is a constant then

$$dx_t/dt = YF(s_r - s_e) - b(x_t - x_g) \qquad 22.23$$

and in the steady state, when the film thickness remains constant, $(dx_t/dt = 0)$ the total amount of biomass $(\tilde{x}_t)$ is given by

$$b(\tilde{x}_t - x_g) = YF(s_r - s_e) \qquad 22.24$$

22.2.5 Limitation by oxygen

In aerobic processes on packed columns oxygen will often be the limiting substrate because of its low solubility. To make a column effective for an aerobic process usually it would be necessary for the liquid medium to be aerated at all stages along the column. It will be assumed that oxygen transfer from gas to liquid medium occurs continuously in each theoretical compartment and that the process rate is limited by oxygen. The oxygen balance in a theoretical compartment is:

rate of increase in = rate of solution − rate of oxygen uptake by
dissolved oxygen   theoretical film of biomass

that is

$$R_0 = K_L a(c_s - c)\frac{V}{n} - x_{gl}q_{O_2} \qquad 22.25$$

Assuming steady-state conditions so that $R_0 = 0$, we have

$$x_{gl}q_{O_2} = \frac{VK_L a}{n}(c_s - c) \qquad 22.26$$

and for the whole column, since all the active theoretical films will be of the same thickness,

$$n x_{gl}q_{O_2} = VK_L a(c_s - c) \qquad 22.27$$

Substituting for x_{gl} by means of Eqn 22.7 and 22.4 we have

$$n = VK_L a(c_s - c)/Kc^{1/2} \qquad 22.28$$

where $K = A(2D\rho q_{O_2})^{1/2}$.

Uptake of oxidized substrate (Δs) is related to the corresponding oxygen uptake (ΔO_2) by the expression $\Delta O_2 = P_0 \Delta s$ or $q_{O_2} = P_0 q$ where P_0 is the 'oxygen demand' constant. The rate of consumption of oxidizable substrate in the whole column is

$$F(s_r - s_e) = n x_{gl}q \qquad 22.29$$

Substituting for x_{gl} from Eqn 22.26 and putting $q_{O_2} = P_0 q$ we obtain

$$s_r - s_e = \frac{VK_L a}{P_0 F}(c_s - c) \qquad 22.30$$

and from Eqn 22.28 and 22.30 we have

$$s_r - s_e = nKc^{1/2}/P_0 F \qquad 22.31$$

22.2.6 Applications

Application of this theory to the design of packed columns for microbial decomposition of substrates leads to the conclusion that, if the medium flow deviates from ideal plug flow, the packed column will become less efficient, since some of the medium exhausted of substrate will wastefully occupy biomass film area. The packed column with plug flow offers the possibility of a series of zones of biomass adapted for different functions such as the sequential utilization of two substrates X, Y where X inhibits or represses utilization of Y. Pirt (1973a) compared the number of theoretical films required for anaerobic and for aerated columns. The data suggest that a packed column could be efficient for anaerobic microbial decomposition but it is severely limited by oxygen solution rate in aerated columns. Quantitative tests of the model are not yet available.

22.3 Growth of submerged pellets of biomass

22.3.1 General features

Mycelial organisms often exist in either of two different forms. One is the homogeneous filamentous form in which the mycelium is homogeneously dispersed in the medium so that all the individual hyphae are exposed to the medium. The second form, termed the 'pellet' or 'stromatic' type, exists as more or less spherical masses in which the hyphae are compacted together (for photographs see Camici et al., 1952). The pellets can grow to a macroscopic size and the packing of the hyphae may be tight enough to exclude substrate permeation except by diffusion. Thus substrate uptake in the pellet could be diffusion limited. On the other hand, mycelial pellets may have an open texture so that medium can enter the pellet by mass flow. The pellet mode of growth is of importance because it commonly occurs in fungal fermentations, also the 'flocs' of sewage microorganisms have similar properties.

The filamentous form of fungal mycelium grows exponentially with constant specific growth rate (μ) until some substrate or an inhibitory product becomes growth-limiting (Pirt & Callow, 1959). Growth of the pellet form deviates from the exponential law and has been found to follow the cube root law

$$M^{1/3} = kt + M_0^{1/3} \qquad\qquad 22.32$$

for many doublings of the biomass (Pirt, 1966; Trinci, 1970). This law would follow if all the growth occurs in an outer shell of the pellet of constant thickness w (Fig. 22.4). According to this model the rate of increase in the pellet radius (r) is given by

$$dr/dt = \mu w \qquad\qquad 22.33$$

where $\mu =$ specific growth rate; hence

$$r = \mu w t + r_0 \qquad\qquad 22.34$$

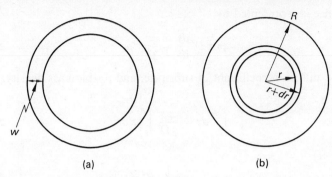

FIG. 22.4   (a) Diagram of cross section through the centre of a biomass pellet with peripheral growth zone of width, w. (b) Cross section through centre of a biomass pellet suspended in a nutrient medium. Substrate concentrations at radius r, $r+dr$, and R are s, $s+ds$ and s_m respectively.

If $M=$ total mass of n pellets (diameters assumed equal) and $\rho=$ density of biomass then $M=\frac{4}{3}\pi r^3\rho n$ and by substituting for r in Eqn 22.34 we obtain Eqn 22.32 where

$$k = (4\pi \rho n/3)^{1/3}\mu w \qquad\qquad 22.35$$

Trinci (1970) found that the radii of submerged pellets of *Aspergillus nidulans* increased in accordance with the linear law (Eqn 22.34). Variation of the specific growth rate by means of temperature showed that $k\propto\mu$ in accordance with Eqn 22.35. The value of w was found to be practically constant at 0·45 mm and therefore independent of the changes in μ. The value of w, it should be noted, is about half the width of the peripheral growth zone of a colony of the same organism growing on agar containing the same medium (Section **23.4.2**). Trinci (1970) also verified that $k\propto n^{1/3}$ as predicted by Eqn 22.35. He also found that, for a period before the cube root law was followed, the pellets of *A. nidulans* followed the exponential growth law. This period, when essentially all of the hyphae contributed to the growth, ended when the pellet diameter was about 4 w.

22.3.2 Limitation by substrate diffusion

If the pellet is so compact that substrates can enter only by diffusion, the size of the pellet, when its growth becomes diffusion limited, is estimated as follows. Consider the substrate balance in the zone from the centre of the pellet to radius r (Fig. 22.4b). We make the simplifying assumption that the metabolic quotient (q) for the growth-limiting substrate is constant and independent of the substrate concentration until it is practically zero, also that there is a steady state such that

Diffusion of substrate into = Uptake of substrate in
the zone of radius, r the zone of radius, r

This balance is represented by

$$4\pi r^2 D' \frac{ds}{dt} = \frac{4}{3}\pi r^3 \rho q \qquad\qquad 22.36$$

where $D' =$ diffusion coefficient of substrate, and $\rho =$ biomass density. Hence we have

$$\int_s^{s_m} ds = \frac{\rho q}{3D'} \int_r^R r\,dr \qquad\qquad 22.37$$

therefore

$$s_m - s = \frac{\rho q}{6D'}(R^2 - r^2) \qquad\qquad 22.38$$

The 'critical' radius of the pellet (R_c) is reached when the growth of the pellet becomes limited by the substrate diffusion, that is when the substrate concentration is zero at the centre of the pellet. Thus $R = R_c$ when $s = 0$ at $r = 0$; putting these values in Eqn 22.38 we obtain

$$R_c = (6D's_m/\rho q)^{1/2} = (6D' Ys_m/\rho\mu)^{1/2} \qquad\qquad 22.39$$

where μ is the specific growth rate and Y is the growth yield. The substrate most likely to be growth-limiting in an aerobic culture is oxygen. With a dissolved oxygen tension of 0·21 atm in the medium, Phillips (1966) found the critical radius of pellets of *Penicillium chrysogenum* to be about 0·1 mm, which is in good agreement with the theoretical value. If glucose (1% w/v) were the growth-limiting substrate, Pirt (1966) estimated the critical radius would be about 1 to 2 mm. A more rigorous and precise theoretical analysis of the diffusion of oxygen into pellets has been given by Aiba & Kobeyashi (1971). In contrast to the low value of the critical radius found for *Penicillium chrysogenum* pellets, Trinci (1970) found that pellets of *Aspergillus nidulans* were not oxygen limited until the radius exceeded 2·5 mm. This suggests that the *Aspergillus* pellets had a much more open texture than the *Penicillium* pellets.

In the centres of dense mould pellets the hyphae autolyse and lose their cytoplasm (Camici *et al.*, 1952). This effect is probably the result of oxygen starvation inside the pellet and indicates that pellet mycelium may become very heterogeneous compared with the filamentous form.

22.3.3 Causes of pellet formation by mycelium

Little is known about the factors which determine whether mycelium will develop in the pellet or filamentous forms. Both the size of the inoculum and the nature of the substrates have been implicated (Camici *et al.*, 1952). These authors found that the pellet form of *Penicillium chrysogenum* was obtained if the conidia in the inoculum were below a critical concentration, which was about $0·3 \times 10^6$/ml in a minimal medium and 10^6/ml in a rich (corn steep liquor) medium. Trinci (1970) found that *Aspergillus nidulans* in minimal medium invariably formed pellets irrespective of the conidia concentration up to $2 \cdot 3 \times 10^6$/ml. The ratio of the number of pellets formed to the number of conidia added was only about 10^{-5}. The surplus conidia aggregated into the pellets. Trinci (1970) suggested that the ability of the conidia to flocculate could be an important factor in pellet formation. The degree of agitation and shear could be important but appears not to have been studied. Pirt & Callow (1959) observed that, in chemostat cultures of *Penicillium chrysogenum*, pellet formation was induced by pH values above $7 \cdot 0$ but not at lower values.

CHAPTER 23

Growth of microbial colonies on the surface of solid medium

23.1 Introduction

Microbial colonization of the surfaces of materials is ubiquitous, and, in the laboratory, growth of colonies of microbes on solid media is one of the basic methods for study of microbial behaviour, consequently the principles of colony growth are of fundamental importance. Since Fawcett (1925) observed that fungal colonies spread outwards at a constant rate, colony growth rates have been used to study factors which influence fungal growth. More recently theoretical and experimental studies have led to understanding of the meaning of colony growth rates. The growth of bacterial colonies is more complex than that of fungal colonies but, under certain conditions, bacterial colonies show 'linear growth' (Pirt, 1967).

23.2 Model of colony growth

Pirt (1967) formulated a model to account quantitatively for colony growth rates. In this model it is supposed that, initially, when a small inoculum of microbes is used to start a colony on nutrient agar, all the biomass contributes equally to growth of the population. Consequently, the population grows at the maximum exponential rate as long as the nutrient concentrations remain much above the saturation constants (K_s values) and no inhibitory condition is developed. The absorption of nutrients from the agar by the colony will cause the formation of a concentration gradient of nutrient within the agar as shown in Fig. 23.1(a). It is assumed that the upward growth of the colony will quickly fall to near zero because of the resistance to nutrient diffusion, as predicted by the theory of substrate diffusion into a plane of tissue (Section 22.1). Eventually the outward growth of the colony is assumed to depend on the exponential growth of a limited peripheral zone of constant width, w, and cross sectional area Δa (Fig. 23.1b). The width, w, of the growing zone is determined by the balance between nutrient consumption and nutrient diffusion into the growing zone. It is assumed that the microbes are incapable of penetrating into the agar gel.

234

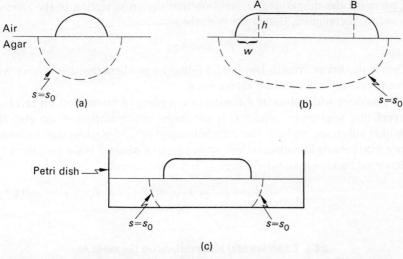

FIG. 23.1   Cross sections of model microbial colonies growing on nutrient agar surfaces. (The diagrams are not to scale.) The broken line shows the depth of the
nutrient concentration gradient in each case; s_0 is the initial nutrient concentration.
(a) Colony during exponential growth. (b) When growth is limited by substrate
diffusion to an annulus of width w. It is assumed that between A and B the growth
rate is limited practically to zero by substrate diffusion. (c) When the depth of the
concentration gradient of a nutrient is limited by the depth of the agar.

Let m = total amount of biomass in the colony and m_g = amount of growing
biomass contained in the peripheral growth zone of width w, depicted in
Fig. 23.1(b). Then the growth of the colony is represented by

$$dm/dt = \mu m_g \qquad\qquad 23.1$$

where μ is the specific growth rate of the organism. It is assumed that w is
small compared with the colony radius, r, so that we can write

$$m_g = 2\pi r\, \Delta a \rho \qquad\qquad 23.2$$

and

$$m = \pi r^2 h \rho \qquad\qquad 23.3$$

where ρ is the density of the biomass. From Eqn 23.3

$$dm/dt = 2\pi r h \rho\, dr/dt \qquad\qquad 23.4$$

Substituting for dm/dt and m_g in Eqn 23.1 and integrating we obtain

$$r = \mu\,\frac{\Delta a}{h}t + r_0 \qquad\qquad 23.5$$

If we make the simplifying assumption that the cross section of the zone of growth is rectangular, that is $\Delta a = wh$, then

$$r = \mu wt + r_0 \qquad\qquad 23.6$$

Thus if the linear growth law holds, Eqn 23.6 predicts that the colony will spread radially at a constant rate, $\mu w = K_r$.

By analogy with substrate diffusion into a plane of tissue (Section **22.1**) we expect that $w \propto (s_0/q)^{1/2}$ where s_0 is the initial concentration of the growth-limiting substrate, and q is the metabolic quotient. Neglecting any maintenance requirement for substrate we can put $q = \mu/Y$ where Y is the growth yield. Hence we find $w = k_1(s_0/\mu)^{1/2}$ where k_1 is a constant and

$$K_r = \mu w = k_1(s_0\mu)^{1/2} \qquad\qquad 23.7$$

23.3 Experimental observations on the mode of growth of bacterial colonies

23.3.1 The linear growth law

Pirt (1967) studied the kinetics of bacterial colony growth to test the model given in Section **23.2**. Accurate and rapid measurement of the colony diameter was made possible by the use of a simple projection microscope (the Shadomaster). The colonies were initiated with about 10^6 bacteria and for a period of about 30 hours after inoculation, the rate of increase in the colony radius was constant, that is the linear growth law was valid; subsequently, the radial growth rate decreased. Similar observations have been reported by Cooper *et al.* (1968) and by Palumbo *et al.* (1971). From the results of Pirt (1967), the width of the growth zone w to account for the radial growth rate during the linear phase is 46 μm for *Escherichia coli* and 25 μm for *Streptococcus faecalis* (originally Pirt (1967) took w to have twice the above values by assuming that the peripheral growth zone had a triangular cross-section). Cooper *et al.* (1968) obtained evidence that the central mass of the bacterial colony does not contribute to the radial spread, in that fine carborundum particles sprinkled on the colony did not move relative to each other. The height of the colonies remained at about 0·5 mm so that the colony may be regarded as a disc of uniform height. In the later stages of growth the colonies collapsed in the centre.

During the period of decelerating rate of radial increase, following the constant rate, Cooper *et al.* (1968) found that the growth corresponded to a constant rate of increase in the area of the colony, since the square of the radius increased linearly with time. This behaviour, Cooper *et al.* suggested, would follow if the model were amended by supposing that w becomes inversely proportional to the radius. If we substitute $\Delta a = bh/r$ in Eqn 23.2 where b is a constant, it follows that

$$r^2 = 2\mu bt + r_0^2 \qquad\qquad 23.8$$

which is termed the area law.

23.3.2 Effect of agar depth on colony growth rate

Pirt (1967) found that the radial growth rate of bacterial colonies reached a maximum when the agar depth exceeded 3·4 mm. Subsequent work (Pirt, unpublished) has shown that increasing the agar depth prolongs the linear growth phase. In fact, on increasing the agar depth to 12 mm, with glucose-limited growth, the maximum rate of radial increase was maintained for 150 hours compared with 35 hours with an agar depth of 3·2 mm. These results suggest that the transition from the linear growth law to the area law is the result of the concentration gradient of growth-limiting substrate extending to the base of the agar layer as shown in Fig. 23.1(c). This conclusion is supported by the calculations of diffusion rates (Cooper et al., 1968), which show that, when the area law applies, the concentration gradient of glucose would reach the base of the Petri dish. From the diagram in Fig. 23.1(c) it may be envisaged that this condition would decrease the width of the peripheral zone which is fully supplied with nutrients.

23.3.3 Effect of limiting substrate concentration

Pirt (1967) showed that if the carbon and energy source (glucose) were made the growth-limiting substrate then, for a limited range of initial glucose concentrations (s_0), the radial growth rate $K_r \propto s_0^{1/2}$ (Fig. 23.2). The upper limits of s_0 for this law to apply to growth of colonies of *Klebsiella aerogenes*, *Escherichia coli* and *Streptococcus faecalis* were 5·0, 2·5 and 1·25 g/l respectively with air at 1 atm pressure. More precisely

$$K_r = k_2(s_0^{1/2} - s_i^{1/2}) \qquad\qquad 23.9$$

where k_2 is a constant and s_i is a threshold concentration, originally called the 'lag' concentration, which must be exceeded for growth to occur. The values of the threshold concentration (s_i) for *K. aerogenes*, *E. coli* and *S. faecalis* were 0·013, 0·090 and 0·005 g/l respectively. The threshold concentration, seems to be peculiar to bacterial colony growth and is not found with filamentous fungi (Section 23.4). The cause of the threshold concentration is not understood; it may be the effect of the maintenance energy requirement or oxygen toxicity (Section 23.3.4). The response of K_r to s_0 could reflect the fact that $w \propto s_0^{1/2}$, as the model suggests (Eqn 23.7).

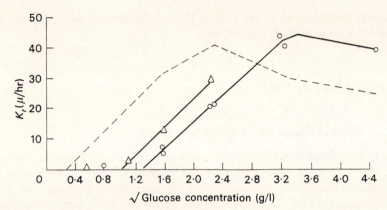

FIG. 23.2   Relation of radial growth rate (K_r) of *Escherichia coli* colonies to the glucose concentration at different oxygen partial pressures. Broken line, growth in air at 1 atm or air + 4·5% CO_2 at 1 atm; △, 95·5% oxygen + 4·5% CO_2 at 1 atm; ○, 100% oxygen at 1 atm. (From Pirt, 1967)

23.3.4 Oxygen limitation and oxygen toxicity

It was found with *E. coli* colonies (Pirt, 1967) that increasing the partial pressure of oxygen (Po_2) from 0·21 to 0·955 atm raised the upper limit of the glucose concentration (s_0) for Eqn 23.9 to be valid from 2·5 to 10·0 g/l. However, the same increase in the oxygen partial pressure also increased the threshold concentration of glucose (s_i) from 0·09 to 1·7 g/l (Fig. 23.2). These results are interpreted to mean that the upper limit of the glucose concentration for Eqn 23.9 to hold is determined by the formation of toxic products such as alcohols and organic acids produced by oxygen-limited metabolism. The increase in the threshold concentration of glucose on increasing the Po_2 may be attributed to oxygen toxicity, since it is surmised that the higher the Po_2, the higher the glucose diffusion rate necessary to meet the oxidation rate which lowers the Po_2 below the inhibitory level.

23.3.5 Influence of specific growth rate

In order to investigate the effect of specific growth rate (μ) Pirt (1967) varied μ both by means of an inhibitor (sulphanilamide) and by temperature. When the growth rate was varied by the inhibitor, the value of the radial growth rate (K_r) was directly proportional to $\mu^{1/2}$ as suggested by the model (Eqn 23.7). However, when μ was varied by means of temperature, the proportionality between K_r and $\mu^{1/2}$ was not generally maintained; the reason for this deviation from the model is unknown.

23.3.6 Instability of colony perimeter

In the later stages of bacterial colony growth, usually after the linear growth law no longer holds, circularity of the colony begins to break down and irregularities begin to appear in the perimeter. Such instability, it was pointed out by Cooper *et al.* (1968), is to be expected if growth is limited by substrate diffusion. This follows because any slight protuberance of the colony at a point on the perimeter would result in the supply of nutrient being increased at that point with consequent accentuation of the protuberance.

23.4 Experimental observations on fungal colony growth

23.4.1 Linear growth law

The rates of growth of colonies of filamentous fungi may be determined either from measurement of the radial spread of colonies in Petri dishes or from the rates of spread of colonies along a layer of agar in a tube (Trinci, 1969). For most purposes the Petri dish method, with a projection microscope to measure colony size, is most convenient. In several important respects the growth of colonies of filamentous fungi differs from that of the unicellular bacteria. Since the fungal hypha grows by extension at the tip, growth of a colony is directed outwards from the centre. The kinetics of growth of colonies of species of *Aspergillus*, *Penicillium*, *Mucor*, *Geotrichum* and other genera were determined by Trinci (1969 & 1971). These results show that after a lag period, which reflects the germination time of the spore inoculum, there is a short period of exponential increase in the colony radius, then the radial growth rate becomes constant and remains so indefinitely. Also the linear growth rate is generally independent of the agar depth; exceptions to this rule are certain morphological mutants (Trinci, 1973). Thus the linear growth law holds indefinitely for fungal colonies, whereas for bacterial colonies it holds only for a limited period.

Also, unlike bacteria, fungal mycelium penetrates into agar gel and, moreover, at a rate which is about equal to the rate of radial spread of the colony on the surface. Thus, below the agar surface, the colony grows in the form of a hemisphere if the agar depth permits (Fig. 23.3). The hyphal density decreases logarithmically with depth of penetration of the mycelium into the agar.

23.4.2 Width of growth zone (w)

This parameter was determined by Trinci (1971) by cutting along a chord of a circular colony so as to cut off hyphal tips of graded lengths, then the growth of the colony was followed by photography. The minimum length of hyphal tip necessary to sustain a growth rate equal to K_r is taken to be w. Thus it was

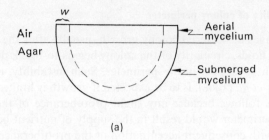

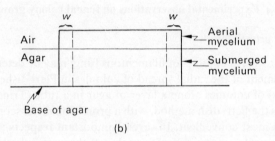

FIG. 23.3 Cross sections of model fungal colonies growing on nutrient agar and penetrating into the agar. (a) Form of colony on deep agar; (b) form of colony on shallow agar. The broken line shows the inner limit of the growth unit, that is the growing part of the hyphal tips.

found that the length w of the growing tip of a hypha, or the 'growth unit' (Caldwell & Trinci, 1973) is characteristic of the species and strain of fungus and varies from about 10 mm for *Neurospora crassa* to 0·4 mm for *Geotrichum lactis*. It is found that the value of w remains constant when the growth rate is varied by temperature or by an inhibitor (cycloheximide) so that K_r stays directly proportional to μ.

The strain-dependent factors which control the length w are obscure, however, the frequency of branching seems to be related to w. The effect of branching is shown by the observation that a mutant of *Aspergillus nidulans*, compared with the parent strain, had a lower value of w and a much greater frequency of branching and, consequently, greater hyphal density (Bainbridge & Trinci, 1969).

Trinci (1970) compared the radial growth rate of a submerged pellet of *A. nidulans* growing in agitated liquid medium with the radial growth rate of a surface colony growing on agar gel with the same medium. It was found that the radial growth rate of the surface colony was about twice that of the submerged pellet. This suggests that the width of the growth zone in the submerged pellet is about one half that in the surface colony. The cause of this difference in the regulation of growth unit length in submerged pellets and aerial hyphae is unknown. These effects might be accounted for if the fungal biomass growth rate is given by $dx/dt = \mu n m_u x$ where both n, the number of hyphal branches per unit of biomass, and m_u the mass of the growth unit may vary.

23.4.3 Effects of substrate concentration

The influence of glucose concentration on the colony growth rate of *Aspergillus nidulans* is depicted in Fig. 23.4. Similar curves with minor modifications

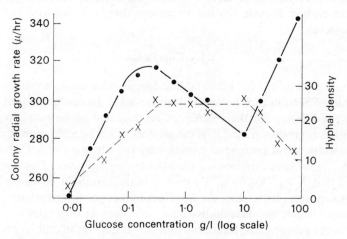

FIG. 23.4 Effect of glucose concentration on radial growth rate (●) and hyphal density (X) of *Aspergillus nidulans* colonies. (Reproduced from Trinci, 1969)

are obtained with other fungi (Trinci, 1969). Unlike bacterial colonies, the relation of fungal colony growth rates to the substrate concentration (s_0) does not conform to the relation $K_r \propto s_0^{1/2}$. Pirt (1973b) concluded that, at the lower substrate concentrations, the relation is of the Monod form, that is

$$K_r = K_{r(max)} s_0 / (s_0 + K_s) \qquad 23.10$$

Since $K_r = \mu w$, it follows that if w is constant, by means of Eqn 23.10, we can calculate the values of the saturation constant (K_s) from measurements of colony growth rates. The values so obtained for fungi are about the same as the values reported in Table 2.1 for other protists.

The values for $K_{r(max)}$ obtained from plots of $1/K_r$ against $1/s_0$ (Pirt, 1973b) are close to the maximum values reached in experiments. At glucose concentrations above the value giving the maximum colony growth rate, K_r may decrease. This decrease was found to be caused by a decrease in w for *Aspergillus nidulans* (Trinci, 1971).

A. nidulans was exceptional in that when the glucose concentration exceeded 10 g/l the K_r started to increase again. This effect was associated with a decrease in the hyphal density. It should be noted that there is frequently some growth at zero glucose concentration on agar, presumably, because some impurity in the agar acts as a carbon source or the agar is attacked and used as a carbon source. Surprisingly, these results indicate that the nutrient concentration gradient which develops around the leading hyphae of a fungal colony on nutrient agar is negligible as appears to be the case for substrate uptake in submerged culture (Section **9.6.2**). The effect of carbon dioxide pressure needs to be investigated since it seems possible that with growth-limiting concentrations of carbon sources the colonies may not generate a sufficient carbon dioxide tension to ensure that it is not growth-limiting (Section **8.10**).

23.5 Conclusion

The linear growth law for bacterial and fungal colonies is accounted for by the model of Section **23.2**, however, the meaning of the constant radial growth rate, K_r, is different for the colonies of bacteria and filamentous fungi. For fungi it has been shown that K_r is the product of the specific growth rate (μ) and the width of the peripheral growth zone (w) where $\mu = \mu_m s_0/(s_0 + K_s)$ and s_0 is the initial concentration of growth-limiting substrate. For bacteria it appears that $K_r \propto \mu^{1/2}(s_0^{1/2} - s_i^{1/2})$ where s_i is the threshold concentration of limiting substrate. By relating the K_r values to the specific growth rate we can apply more widely the quantitative study of colony growth rates to study of the influence of nutrients, inhibitors and stimulators on microbial growth.

CHAPTER 24

Mathematical models of biomass autosynthesis

24.1 Introduction

Since growth of living cells is a process of autosynthesis, growth must conform to the principles of chemical autosynthesis. It is therefore relevant to look at a model of autosynthesis to see whether it can add to our understanding of the process of growth and enable us to deduce and predict any properties of biomass. The models of autosynthesis, first developed by Hinshelwood (Dean & Hinshelwood, 1966), are discussed below. These models are enormously simplified, but nevertheless they give valuable indications of the behaviour of growing biomass, particularly for steady-state conditions which can be achieved in chemostat culture. However transient conditions, which are typical of batch culture, introduce complexities which give little scope for application of the models.

24.2. Interdependent syntheses

24.2.1 Cycles of interdependent syntheses

Autosynthesis of biomass is the result of interdependent syntheses of which the simplest case is a two-component system (Fig. 24.1a). An arrow, for instance, in $X \leftarrow Y$ means that the synthesis of component X depends on the action of component Y; it does not necessarily mean that X is the product of Y. Examples of cycles of interdependent syntheses which may be abstracted from biomass synthesis are shown in Fig. 24.2. The two-component system consists of synthesis of DNA which depends on DNA polymerase, and synthesis of DNA polymerase, which depends on DNA for the supply of messenger RNA.

If we represent the total biomass by x during exponential growth in a constant environment we have

$$x = x_0 e^{\mu t} \qquad 24.1$$

where μ is the specific growth rate. It follows that each component of the system, if it forms a constant fraction of the biomass, must increase in amount with constant specific growth rate, μ.

243

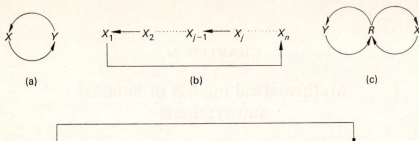

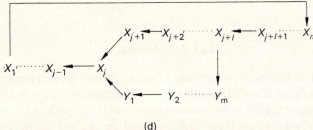

(d)

FIG. 24.1 Diagrams of autosynthetic systems. The X and Y symbols represent structures in the cell, such as enzymes, DNA and ribosomal RNA. An arrow between two structures, for instance, $X \leftarrow Y$, means that the synthesis of X is dependent on the action of Y. All the systems are composed of cycles of interdependent syntheses. (a) Two-component cycle; (b) n component structures in a cycle; (c) and (d) branched cycles.

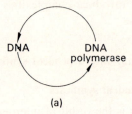

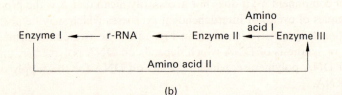

(b)

FIG. 24.2 Examples of interdependent syntheses. (a) Two-component system; DNA synthesis depends on DNA polymerase, and DNA polymerase synthesis depends on DNA for the supply of messenger (m-)RNA. (b) Four-component system. Synthesis of enzyme I depends on ribosomal (r-)RNA; synthesis of r-RNA depends on enzyme II; synthesis of enzyme II depends on enzyme III because it produces amino acid I which is required for synthesis of enzyme II; synthesis of enzyme III depends on enzyme I because it produces amino acid II required in synthesis of enzyme III.

The interdependences of the rates of synthesis of X and Y in the two-component system (Fig. 24.1a) are expressed by the equations

$$dX/dt = \alpha Y \qquad\qquad 24.2$$

and

$$dY/dt = \beta X \qquad\qquad 24.3$$

where X and Y are used to represent the amounts of the components and α and β are constants, the value of which will depend on the environmental conditions including the supply of diffusible intermediates, for instance the mRNA in Fig. 24.2(a) and amino acids in Fig. 24.2(b). The solutions of Eqn 24.2 and 24.3 give

$$X = \tfrac{1}{2}\left(X_0 + \frac{\alpha}{k} Y_0\right) e^{kt} + \tfrac{1}{2}\left(X_0 - \frac{\alpha}{k} Y_0\right) e^{-kt} \qquad\qquad 24.4$$

and

$$Y = \tfrac{1}{2}\left(Y_0 + \frac{k}{\alpha} X_0\right) e^{kt} + \tfrac{1}{2}\left(Y_0 - \frac{k}{\alpha} X_0\right) e^{-kt} \qquad\qquad 24.5$$

where $\alpha\beta = k^2$ and X_0 and Y_0 are the amounts of X and Y at zero time. When t becomes large the term in e^{-kt} vanishes and the ratio of $X : Y$ becomes

$$\frac{X}{Y} = \left(X_0 + \frac{\alpha}{k} Y_0\right) \Big/ \left(Y_0 + \frac{k}{\alpha} X_0\right) = \frac{\alpha}{k} \qquad\qquad 24.6$$

that is the ratio of $X : Y$ settles down to a constant value. If some of the system is now isolated and used to start a new one in which $X_0/Y_0 = \alpha/k$, then Eqn 24.4 and 24.5 reduce to

$$X = X_0 e^{kt} \quad\text{and}\quad Y = Y_0 e^{kt} \qquad\qquad 24.7$$

Hence the model accounts for exponential growth and since all the components in the system must have the same specific growth rate as the biomass, it follows that $k = \mu$.

The result is found to be general in that it applies if there are n components in the system (Fig. 24.1b) where $dX_1/dt = \alpha_1 X_2$, $dX_2/dt = \alpha_2 X_3$, ... $dX_j/dt = \alpha_j X_{j+1}$ and finally $dX_n/dt = \alpha_n/X_1$ where α_1, α_2 etc. are constants. For the steady state we can substitute $dX_1/dt = \mu X_1$, $dX_2/dt = \mu X_2$ etc., from which it follows that the ratio of the amounts of X_j and X_{j+1} will have the constant value, $X_j/X_{j+1} = \alpha_j/\mu$, where

$$\mu^n = \alpha_1\alpha_2\ldots\alpha_n \qquad\qquad 24.8$$

When there are more than two structural components in the system the solutions of the equations for the transient state contain terms in $\sin kt$ (Hinshelwood, 1952; Dean, 1969), which shows that the proportions of the components will oscillate before they settle down to the final steady-state values.

24.2.2 Branched essential cycles

The synthesis of complex structural components such as enzymes will, in general, be the result of branched cycles of interdependent syntheses. The simplest case will be the three-component system illustrated in Fig. 24.1(c). Branched cycles can be of two kinds, those in which each cycle is essential and those in which the cycles are alternative systems of syntheses (Section 24.5). As an example of essential cycles we could have in Fig. 24.1(c), R = amount of ribosomal RNA, X and Y are amounts of enzymes providing nucleotides for the synthesis of RNA. In the system of essential cycles represented in Fig. 24.1(c), the R–X cycle is modelled by $dX/dt = \alpha R$ and $dR/dt = \beta X$. In the steady state when $dX/dt = \mu X$ and $dR/dt = \mu R$ we find

$$X/R = \alpha/\mu \quad \text{and} \quad \alpha\beta = \mu^2 \qquad 24.9$$

For the other essential cycle modelled by $dY/dt = aR$ and $dR/dt = bY$, in the steady state we shall have

$$Y/R = a/\mu \quad \text{and} \quad ab = \mu^2 \qquad 24.10$$

Hence $\mu^2 = ab = \alpha\beta$. It also follows that $X/Y = \alpha/a$, hence although the ratio of $X:R$ and of $Y:R$ can vary with the specific growth rate, the ratio of $X:Y$ is independent of μ.

24.3 Automatic adjustment to environmental change

The effect of a change in the environment on an autosynthetic process is deduced as follows (Dean & Hinshelwood, 1966, p. 130). In the two-component system represented in Fig. 24.1(a), let us suppose that structure X is an enzyme that provides a diffusible metabolite necessary for synthesis of the second enzyme, Y. Let the total amount of enzyme in the population of cells be X, and if n is the number of growing cells, the amount per cell is X/n. Similarly for the second enzyme the amount per cell is Y/n. Let c be the concentration of the diffusible metabolite in the cell then, for the net rate of accumulation of the metabolite in a single cell, we have

$$dc/dt = A(X/n) - B(Y/n)c - Cc \qquad 24.11$$

where A, B and C are constants. In Eqn 24.11 the first term on the right represents the rate of production of the metabolite, the second term the rate of consumption, and the third term the losses of the intermediate by decomposition or other mechanisms. We make the simplifying assumption that there is a steady state so that the production of the intermediate just balances the consumption and losses, that is $dc/dt = 0$. Also in the steady state we have

$$dX/dt = k_1 X \qquad 24.12$$

and

$$dY/dt = k_2 Yc \qquad 24.13$$

where k_1 and k_2 are constants. Equation 24.13 expresses the condition that, the synthesis of Y depends on both its own amount and the amount of the diffusible metabolite. We further assume that the division of the cell is provoked when the concentration, Y reaches a critical value given by $\beta = Y/n$ where β is a constant. Substituting for n in Eqn 24.11 we obtain

$$c = \alpha X/Y \qquad\qquad 24.14$$

where α is a new constant. From Eqn 24.13 and 24.14 we obtain

$$d Y/dt = k_2\alpha X \qquad\qquad 24.15$$

Let $v = X/Y$ then on differentiation we obtain

$$dv/dt = \frac{Y(d X/dt) - X(d Y/dt)}{Y^2} \qquad\qquad 24.16$$

that is

$$dv/dt = v(k_1 - k_2\alpha v) \qquad\qquad 24.17$$

In the steady state if $\mu =$ specific growth rate of the biomass, we have from Eqn 24.12, 24.13 and 24.15

$$\mu = k_1 = k_2 c = k_2\alpha v \qquad\qquad 24.18$$

which makes $dv/dt = 0$. Now suppose some agent such as a toxic substance decreases the concentration (c) of the metabolite. This means that αv is decreased (from Eqn 24.14) and consequently, from Eqn 24.17, it is apparent that dv/dt becomes positive. Thus the ratio $v = X/Y$ increases and will go on doing so until once again $k_2\alpha v = k_1 = \mu$, that is the growth rate is restored with a new steady-state value of X/Y. Hence the model predicts that an autosynthetic system such as biomass will react to a change in the conditions which affects growth rate by changing the proportions of the structural components so as to restore the growth rate.

24.4 Magnitude of an environmental effect on growth rate

The model indicates how much the specific growth rate (μ) could be affected by a change in the amount of a single enzyme or other essential structure in a cycle of interdependences. From Eqn 24.8 we have $\mu = (\alpha_1\alpha_2\ldots\alpha_n)^{1/n}$. The values of the α coefficients will reflect the supply of metabolic intermediates. Let us suppose that some inhibitor or other environmental effect decreases the supply of a metabolite so that the α_1 coefficient is decreased to a fifth of its original value. The ratio of the new value of μ to the original value, it follows from Eqn 24.8, will be $(0\cdot2)^{1/n}$. Thus if the number of structures in the cycle is 20, the ratio of the new value of μ to the old becomes $(0\cdot2)^{0\cdot05} = 0\cdot91$, or the value of μ is reduced by only 9%. The ratio of the amounts of structures X_1 and X_2 would change from α_1/μ to $0\cdot2\alpha_1/\mu$. Thus the model indicates that the proportions of enzymes in the biomass may substantially change without much affecting the growth rate.

24.5 Alternative cycles of interdependent syntheses

24.5.1 Branched alternative cycles

In the system of branched cycles of interdependent syntheses shown in Fig. 24.1(c) each cycle could be an alternative to the other. This system is modelled as $dX/dt=\alpha R$, $dY/dt=aR$ and $dR/dt=\beta X+bY$ where α, a, β and b are constants. In this system α and a determine competing or alternative dependences. In the final steady state we find that $X/R=\alpha/\mu$ and $Y/R=a/\mu$ and the specific growth rate, μ, is given by

$$\mu^2 = \alpha\beta+ab \qquad\qquad 24.19$$

(Hinshelwood, 1953). Hence this system also accords with the general result that the ratios of the amounts of the structures will settle down to constant values.

24.5.2 The network theorem

Another important deduction from the model concerns the development of alternative enzyme pathways (Dean & Hinshelwood, 1966, p. 111). With the interdependences illustrated in Fig. 24.1(d) it is supposed that synthesis of enzyme X_j requires a metabolite C formed by either enzyme X_{j+1} or by enzyme Y_1, which is on a different branch of the system. Let the local concentration of the metabolite be c, then we should have for the steady state

$$dc/dt = \beta X_{j+1}+\gamma Y_1-\delta c = 0 \qquad\qquad 24.20$$

where β, γ and δ are constants. The first term on the right represents the production of c due to X_{j+1}, the second term represents the production due to Y_1 and the third term represents the consumption of C in synthesis of X_j plus any losses of C. It follows that

$$c = (\beta/\delta)X_{j+1}+(\gamma/\delta)Y_1 \qquad\qquad 24.21$$

and if dX_j/dt is proportional to c we have

$$dX_j/dt = \alpha_j X_{j+1}+\beta_j Y_1 \qquad\qquad 24.22$$

where α_j and β_j are new constants. From the equations for the interdependences in the network given in Fig. 24.1(d), that is $dX_1/dt=\alpha_1 X_2$ etc. to

$dX_n/dt=\alpha_n X_1$ and $dY_1/dt=b_1 Y_2$ etc. to $dY_m/dt=b_m X_{j+1}$, and Eqn 24.22, it follows that in the steady state where we can substitute $dX_1/dt=\mu X_1$ etc.

$$\mu^n = \alpha_1\alpha_2\ldots\alpha_{j-1}(\alpha_j\ldots\alpha_{j+l-1}+\beta_j b_1\ldots b_m\mu^{l-m-1})\alpha_{j+l}\ldots\alpha_n \qquad 24.23$$

If we simply have Y, spanning the gap between X_j and X_{j+2} then Eqn 24.23 becomes

$$\mu^n = \alpha_1\alpha_2\ldots\alpha_{j-1}(\alpha_j\alpha_{j+1}+\beta_j b_1)\alpha_{j+2}\ldots\alpha_n \qquad 24.24$$

It follows that any decrease in the α_j or α_{j+1} values can be compensated by an increase in β_j or b_1. The effect of a change in the environment which affects only one of the branches can be deduced as follows. In the steady state we have $dX_j/dt=\mu X_j$, then from Eqn 24.22 we obtain

$$\mu = \alpha_j\frac{X_{j+1}}{X_j}+\beta_j\frac{Y_1}{X_j} \qquad 24.25$$

With one particular set of environmental conditions the term $\beta_j Y_1/X_j$ may be very small so that the effect of the X_{j+1} component is dominant over that of Y_1. A change in the conditions, which decreases the α_j coefficient to a low level, would initially decrease the biomass growth rate but then the production of Y_1 would be stimulated according to the automatic mechanism discussed in Section 24.3. Thus gradually the growth rate could be restored to its former level with the effect of the Y_1 component dominant over that of X_{j+1}. This argument, termed the *network theorem* (Dean & Hinshelwood, 1966, p. 113), therefore predicts behaviour which is analogous to enzyme repression and induction.

24.6 Conclusion

From the basic concept that the growth of living matter depends upon closed cycles of interdependent syntheses the Hinshelwood kinetics predict that, in a constant environment, when the biomass is in a steady state, the proportions of the various constituents settle down to constant values. Another salient feature of the kinetics of autosynthesis is that the process is self-regulating. This self-regulation causes the proportions of enzymes, nucleic acids and other essential structural elements to vary in response to environmental changes so as to maintain the biomass growth rate, and conversely the biomass composition must vary with the growth rate. Experiments show that the RNA content and enzyme activities of biomass vary with the growth rate (Pirt, 1969). There have been few studies on the effects of varying nutritional and physical conditions on enzyme activities when the growth rate is held constant by means of chemostat culture. Decreasing the dissolved oxygen tension greatly alters the activities of some enzymes (Section 11.5). Unpublished work in the author's laboratory (Watts Evans & Pirt) shows that the amount of certain glycolytic enzymes in chemostat steady states of *Lactobacillus* vary several fold with change in pH value and twofold with change in temperature. Some evidence from studies on chemostat cultures show that inhibitors alter the amounts of enzymes in cells (Section 17.1). These inhibitor studies also show that, immediately following the addition of an inhibitor, oscillations in the growth rate can occur as predicted by the transient state kinetics. Thus the dynamic state of biomass during growth, predicted by the model of autosynthesis, is qualitatively in agreement with the experimental observations.

Dean & Hinshelwood (1966, p. 113) commented that the model 'may well oversimplify the nature of the mutual dependences. This, however, is very unlikely to have led to the prediction of possibilities which do not exist, and, indeed, the actual behaviour of living matter is probably more and not less subtly adaptive than we have foreseen.' The repressor mechanisms and feedback or feedforward inhibition or activation may be regarded as *microregulation* evolved to refine and make more sensitive the inherent *macroregulation* of autosynthesis of biomass.

CHAPTER 25

Abstract and conclusion

Growth and functions of biomass and the influence of environment upon them are expressed quantitatively in terms of the parameters: lag period, specific growth rate, growth yield, metabolic quotients for substrate and product, saturation constant (K_s) and biomass concentration. Hyperbolic (Michaelis–Menten, or Monod) type relations between metabolic quotients and substrate concentration are to be expected.

The determination of biomass is a key method in the quantitative study of microbe cultivation. A wide variety of methods is available for the determination of biomass in minimal, homogeneous liquid media, but with increasing complexity of the medium and with mixtures of microbial species the choice of method may be limited severely. Other methods of biomass determination need to be developed. The biomass may be estimated by any parameter which can conveniently be related to biomass.

Closed systems of culture such as simple batch culture are always in a transient state in which process rates tend towards zero. In simple batch cultures two states are successively dominant. The first is the exponential growth phase in which the organism is supplied with excess of substrate and growth occurs at the maximum rate. The second is the stationary phase which is characterized by zero growth rate and reorganization or degradation of the biomass structure.

Substrate-limited growth, which is probably more commonly required than growth with excess substrate, is achieved by some form of continuous feed. In a simple batch culture, substrate-limited growth normally occurs only briefly at the end of the last doubling of the biomass, however, the period of substrate-limited growth can be prolonged by means of a substrate feed. Such a culture is called a fed batch culture. When a part of a fed batch culture is withdrawn periodically it becomes a repeated fed batch culture, which reproduces periodically the transient conditions between two different specific growth rates.

Dialysis culture is a special form of fed batch culture which makes possible the concentration of biomass and non-diffusible products formed from dilute media. Also it is a means of lowering the concentration of a non-diffusible inhibitory product in a culture.

In chemostat culture the full range of environmental conditions can be realized with any desired growth rate. Also any particular condition can be maintained indefinitely. Thus the unique features of the chemostat are that it enables growth rate to be varied with a constant environment, or vice versa, the environment can be varied whilst the growth rate is maintained constant.

The turbidostat is an elaboration on the chemostat which is advantageous for maintaining a constant biomass where simple chemostat control is unstable, for instance, near the maximum growth rate. Concentration of biomass in a chemostat by biomass feedback is advantageous for increasing process rates with dilute media and makes the culture more stable under conditions of shock loading with inhibitors. With two chemostats in series, the second stage may be used to maintain a constant biomass at zero growth rate, also to obtain a stable culture at maximum specific growth rate.

Plug-flow culture simulates simple batch culture conditions in a continuous process, thus the various phases are spatially separated. True plug-flow culture is technically more difficult to achieve than chemostat culture. However, a good approximation to plug-flow culture can be achieved by a number of chemostats in series.

When a constant proportion of the cells produced in a culture are non-viable, logarithmic growth is still possible. However, the environmental conditions will be affected if the dead cells autolyse or release their cell constituents.

Energy is required for growth and maintenance of biomass. The amount of energy required can be predicted from estimation of the ATP required for synthesis of biomass, expressed as $1/Y_{ATP}^G$, and for maintenance, m_{ATP}. The value of Y_{ATP}^G depends on the nutrition of the organism. For heterotrophs, with carbohydrate as the energy source, Y_{ATP}^G is about 25 g dry biomass/mole ATP; for autotrophs the Y_{ATP}^G is about 5 g dry biomass/mole ATP. The value of the maintenance energy may vary many fold with environmental conditions and thus cause considerable variation in the overall value of Y_{ATP}.

The metabolic fates of carbon and energy sources are affected by growth rate and by environmental conditions; major effects are caused by oxygen tension, pH value, temperature, water activity or osmolality and nutrition. Little is known about the interactions in the metabolism of mixed carbon sources apart from the repression of enzyme synthesis and inhibition associated with diauxie. However, mixed carbon sources can be utilized simultaneously in carbon-limited chemostat cultures. There has been little systematic study of the effect of carbon dioxide tension on microbial growth. The K_s value for carbon dioxide corresponds to a value about equal to that of carbon dioxide in air at 1 atm, hence production of carbon dioxide by the organism is probably essential to maintain the optimum value. The uptake of insoluble hydrocarbons probably occurs mostly by direct oil to biomass contact.

The maximum rate of supply of oxygen to cultures is usually governed by three basic factors, which are the oxygen tension in the gas phase, the gas–liquid interfacial area and the resistance to oxygen transfer across the gas–liquid interface. The resistance to oxygen transfer of the stationary liquid film surrounding the biomass is, for most purposes, considered to be negligible.

In aerobic cultures dissolved oxygen tension is accurately measured by means of the oxygen electrode. Redox potential is a means of measuring dissolved oxygen tension but, because of its logarithmic response, it is much less sensitive than the oxygen electrode except at extremely low values of oxygen tension close to anaerobic conditions.

Agitation is necessary both for mixing and aeration of cultures. Mixing is essential in chemostat cultures to ensure that the incoming medium is well mixed with the culture. The vortex aeration system is more convenient than the baffled system on the small scale (up to about 10 litres). The baffled system is used to obtain high oxygen transfer rates with minimum power consumption.

Oxygen-limited growth of aerobic organisms begins to occur with a dissolved oxygen tension about 10 mmHg, the actual value will be a function of the specific growth rate. The K_s value of oxygen for growing cultures lies in the range 10^{-5} to 10^{-6} M. The respiration rate of non-growing microbial cells is greatly affected by the previous growth conditions. The transition between aerobic and anaerobic metabolism in growing cultures of facultative anaerobes occurs at about 5 mmHg. In the absence of oxygen, ferricyanide and some other electron acceptors may substitute for oxygen.

The contents of respiratory enzymes and other enzymes in biomass are affected by a wider range of dissolved oxygen tension than is the growth rate. For all species of organisms oxygen may be regarded as an inhibitory substrate. With common aerobes the inhibitory oxygen tension is less than 1 atm. Anaerobes are defined as organisms, which have their growth inhibited by oxygen at any tension. The less strict anaerobes are probably organisms which have a mechanism for consuming oxygen and so lowering the tension to a tolerated level.

Historically, the qualitative aspects of microbial nutrition have received the most attention. The quantitative requirements for nutrients, expressed in terms of the 'growth yield', are now assuming major importance. The growth yield can be much influenced by the other nutrients supplied, the physico-chemical environment and the specific growth rate. When the requirements for amino acids, vitamins and other growth factors are complex, interactions between the different nutrient effects often occur. The use of defined media is still remarkably limited. The failure to achieve growth of many species of organism, or maximum productivity in defined media, reflects our ignorance of both qualitative and quantitative requirements for nutrients. Knowledge of the roles and quantitative requirements for nutrients is being greatly extended by the study of substrate-limited growth. Least is known about the effects of growth limited by the supply of organic growth factors or trace elements.

Trace metal ion concentrations in media are almost inevitably subject to modification and buffering by chelating agents. Control by metal chelation needs systematic study.

The design of culture media is based on a synthesis of knowledge of all the qualitative and quantitative requirements for nutrients, the interactions between substrates, the effects of inhibitors and culture products, the physical conditions and medium stability.

Temperature affects the nutrient requirements, the biomass composition and the nature of the metabolism in cultures. The effects of temperature on growth rate over most of its range can be predicted in terms of the 'activation' energy for growth. This relation breaks down near to both the upper and lower temperature limits for growth. Above the optimum temperature, cell degradation probably becomes dominant over the growth process. Near the lower temperature limit the regulation of metabolism may fail.

The intracellular and extracellular pH values are probably different. The medium pH affects both the biomass composition and the nature of the metabolism, but the molecular basis of these effects is not understood. Possibly the proton motive force generated across the plasma membrane in chemiosmosis is affected by the medium pH value.

The water availability in a culture is expressed either in terms of the water activity or as the osmolality. Water activity is a convenient parameter for the more xerophilic conditions, however, for many common organisms the optimum water activity is greater than 0.99, and for these the parameter osmolality is preferred to water activity. The water activity affects the specific growth rate, biomass composition and metabolism.

The total amount of a metabolic product formed in a fermentation depends on four basic factors: biomass concentration, specific rate of product formation (q_p), duration of the synthetic activity, and decomposition rate of the product. The q_p value is determined by the genome and by the substrates, inhibitor action and other environmental factors. The relations between the q_p value and the specific growth rate fall into three classes: (i) the q_p increases with the growth rate, which is characteristic of primary metabolism, (ii) the q_p is independent of the specific growth rate above a critical value, and (iii) the q_p may vary in a complex way with the growth rate. Factors controlling the duration of the synthetic activity in biomass at zero growth rate are little understood.

The effect of a growth inhibitor in a simple batch culture is manifest by a reduction in the maximum growth rate. In chemostat culture, growth inhibitors will affect the relation of the steady-state biomass to the dilution rate in a characteristic way dependent on whether the inhibition is competitive or non-competitive with substrate uptake. In the short term after its addition a growth inhibitor may cause oscillations in the growth rate. Any growth-inhibitory product in a batch culture will progressively decrease the growth rate. However, in a chemostat culture it is predicted that an inhibitory product may affect the relation between steady-state biomass concentration and dilution rate in a characteristic way depending on whether the product inhibition is competitive or non-competitive and the relation between q_p and the specific growth rate, μ.

Substrate inhibition of growth causes a characteristic pattern of change in the steady-state biomass with dilution rate. The instability of a chemostat culture with a growth-limiting, inhibitory substrate can be overcome in a second stage chemostat.

The possibility of increasing maximum specific growth rates by substances which act as activators needs to be investigated.

There is evidence that there is a finite minimum growth rate at least for bacterial biomass. This may be obscured by the presence of both growing and dormant (non-growing) cells in the culture. Both bacteria and fungi at very low growth rates have drastic changes induced in their structure and function. In some cases these changes lead to reorganization of the biomasses into spore forms. If the growth is abruptly stopped by the withdrawal of the energy supply, autolysis is favoured, however, if the maintenance energy is supplied, autolysis is inhibited.

The lag phase preceding growth of a culture may be due to a variety of causes. The so-called inoculum effect which results from use of an inoculum taken from a stationary phase culture is of physiological interest since there is selective stimulation of some enzymic processes during the lag. 'Early lag' is the result of a lack of intermediary metabolites. The lag period, under some conditions, is proportional to the doubling time of the biomass during logarithmic growth in the same medium.

So far there have been few systematic studies of the interactions between species in mixed cultures. Theoretical models indicate that the interactions between species can best be determined in chemostat culture rather than simple batch culture. If there is uncontrolled competition for the same limiting substrate, then only that species with the highest affinity for the substrate can remain in a homogeneous open system. The competition may be controlled by the presence of inhibitors or activators, if these substances are produced by the appropriate species, so that in effect the species cooperate to maintain each other in the population. Stable mixed populations also occur if there are different growth-limiting substrates for each species. When two species are not in competition for the same growth-limiting substrate one may assist the other by the supply of an essential nutrient or the removal of an inhibitor. The presence of a predator species can decrease the degree of utilization of the growth-limiting substrate by the prey species. With a prey–predator culture a two-stage process is necessary for complete conversion of growth-limiting substrate through prey into predator biomass. A predator may stabilize a population of two other competing organisms.

The packed column with microbes attached is important in soil microbiology and in the trickling filter method of effluent purification. The performance of such a culture system is predicted in terms of the number of 'theoretical films' of biomass each with an area equal to that of a cross section of the column and with thickness equal to the depth of diffusion of the growth-limiting substrate, which probably will be oxygen in aerobic processes.

Growth of biomass in the pellet form is characteristic of many filamentous fungi. The pellets become heterogeneous with growth limited by nutrient diffusion to a peripheral layer when the pellet size reaches a critical value.

The outward growth of a surface colony of microbes is attributed to growth in a peripheral zone of constant width, w. Fungal colonies spread outwards at a constant rate which is the product of the specific growth rate (μ) and the width w of the growing zone. The width w is considered to be the length of the growing tip of a hypha, that is the 'growth unit'. Also bacterial colonies spread outwards at a constant rate for a period which, on nutrient agar plates, is determined by the depth of the agar. After the linear growth period of a bacterial colony is over, the square of the radius of a colony, or the colony area, increases at a constant rate.

The relation of the linear growth rate of a fungal colony to the initial concentration of growth-limiting substrate is expressed by the Monod hyperbolic relation at low substrate concentration. During the constant linear growth period of a bacterial colony the linear growth rate is proportional to the square root of the initial concentration of growth-limiting substrate and possibly to the square root of the specific growth rate.

The simplest model of growth of biomass is based on an autosynthetic system consisting of two structures the syntheses of which are interdependent. Elaboration of the model presents a cell as a closed network of structures in which the syntheses are interdependent. The model predicts that the amounts of enzymes and other structures in the system will settle down to constant ratios characteristic of a given environment and which maximize the growth rate in the environment. Further the model predicts that a change in the environment will automatically alter the proportions of the enzymes so as to maintain the maximum growth rate. Where there are alternative enzymes to supply a given metabolite for synthesis of another enzyme, it is predicted that a change in conditions which permits one synthetic cycle to work and not the other will automatically stimulate synthesis of the enzyme which can function, and depress the synthesis of the other. This provides a kinetic basis for the curative properties of nature! This selfregulation inherent in an autosynthetic system probably is sluggish and needs the well-known enzyme repressors and feed-back inhibition mechanisms to make the regulation more sensitive and rapid.

The science of microbial cultivation is still at its beginning and all the principles considered need further testing and development. Since the number of microbial and tissue cell species cultivated under controlled conditions is only a minute proportion of the whole, most of the diversity of biomass behaviour in pure and mixed cultures remains to be elucidated and formulated by general principles.

APPENDIX

Symbols and abbreviations

The symbols most frequently used in the text have the following meanings unless it is otherwise stated.

a	Specific maintenance rate ($= m Y_{EG}$)	m_o	Osmolality or tonicity (a 1 osmolal solution exerts an osmotic pressure of 22·4 atm at 0°C)
a_w	Water activity		
atm	Atmospheres (pressure)		
ATP	Adenosine triphosphate	mm	Millimetres
cP	Centipoises (viscosity)	mmHg	Millimetres mercury (pressure)
D	Dilution rate ($= F/V$)	p	Product concentration
D_c	Critical dilution rate (when μ is at maximum)	P	Total amount of product in culture (Vp)
D'	Diffusion coefficient	P_0	Oxygen demand constant, moles oxygen consumed/mole of oxidizable substrate consumed
DNA	Deoxyribonucleic acid		
DOC	Dissolved oxygen concentration (g/l)		
		pM	$-\log m_e$ where m_e is the molarity of a metal ion
DOT	Dissolved oxygen tension (mmHg)		
E_h	Oxidation–reduction or redox potential (Eqn 9.4)	ppm	Parts per million
		q_p, q_s	Metabolic quotients, $1/x . dp/dt$ and $1/x . ds/dt$ respectively, abbreviated to q where indicated
Eqn	Equation		
F	Medium flow rate		
H	Henry's constant (Eqn 9.2)	RNA	Ribonucleic acid
i	Ionic strength (Eqn 15.1)	s	Substrate concentration
K_a	Activator constant (Eqn 17.44)	S	Total amount of substrate in culture (Vs)
K_i	Inhibitor constant defined in Section 17.2.1		
		s_r	Substrate concentration in medium feed
$K_L a$	Gas-transfer coefficient (Eqn 9.13); $a =$ gas–liquid interfacial area; $1/K_L =$ resistance to gas diffusion		
		t	Time
		T	Temperature
		t_d	Doubling time or mean generation time of biomass
K_s	Saturation constant (Eqn 2.21)		
K_r	Colony radial growth rate	t_r	Mean residence time or replacement time of a continuous flow culture ($= V/F$)
L	Lag period before growth		
ln	Logarithm to base e		
log	Logarithm to base 10	V	Culture volume
m	Maintenance coefficient (Section 8.3.1), also molality (moles solute/kg water)	vol.	Volume
		w/v	Weight per volume
		w/w	Weight per weight
M	Molarity (moles/litre)	x	Biomass concentration

258

Y_{ATP} Overall ATP yield (g dry biomass produced/mole ATP)

Y_{ATP}^G True ATP yield, that is excluding effect of maintenance energy

Y_c $Y_{x/s}$ where the substrate is the carbon (not energy) source (Eqn 8.4)

Y_E Overall growth yield, $Y_{x/s}$ where the substrate is the energy source

Y_{EG} True growth yield, $Y_{x/s}$ where the substrate is the energy source and $m = 0$

$Y_{p/s}$ Product yield $(-dp/ds)$, abbreviated to Y_p where indicated

$Y_{p/x}$ Product yield (dp/dx), abbreviated to Y_p where indicated

$Y_{x/s}$ Growth yield $(-dx/ds)$, abbreviated to Y_x or Y where indicated

μ Specific growth rate, $1/x . dx/dt$

μ_m Maximum specific growth rate defined by Eqn 2.21

μm Microns

π Osmotic pressure (atm)

ϕ Osmotic coefficient (Eqn 15.8)

∞ Infinity

$\propto$ is proportional to

$>$ is greater than

$\gg$ is much greater than

$<$ is less than

$\ll$ is much less than

$\approx$ approximately equals

$\sim$ (over a symbol) denotes steady-state value

References

AIBA S., HUMPHREY A. E. & MILLIS N. F. (1965) *Biochemical Engineering*. Academic Press, New York

AIBA S. & KOBAYASHI K. (1971) *Biotech. Bioeng.* **13**, 583

AIBA S., NAGATANI M. & FURUSE M. (1967) *J. Ferm. Technol.* (*Japan*) **45**, 475

AIBA S., SHODA M. & NAGATANI M. (1968) *Biotech. Bioeng.* **10**, 845

AIDA K. (1972) In *The Microbial Production of Amino Acids*, ed. Yamada K. *et al.*, p. xiii. John Wiley, New York

ANAGNOSTOPOULOS G. D. (1973) *J. gen. Microbiol.* **77**, 233

ARNON D. I. (1938) *Amer. J. Bot.* **25**, 322

AUBEL E., ROSENBERG A. J. & GRUNBERG M. (1946) *Helv. Chim. Acta* **29**, 1267

BABIJ T., MOSS F. J. & RALPH B. J. (1969) *Biotech. Bioeng.* **11**, 593

BAIDYA T. K. N., WEBB F. C. & LILLY M. D. (1967) *Biotech. Bioeng.* **9**, 195

BAIL O. (1929) *Z.f. Immunitätsforschung* **60**, 1

BAINBRIDGE B. W., BULL A. T., PIRT S. J., ROWLEY B. I. & TRINCI A. P. J. (1971) *Trans. Br. mycol. Soc.* **56**, 371

BAINBRIDGE B. W. & TRINCI A. P. J. (1969) *Trans. Br. mycol. Soc.* **53**, 473

BANDYOPADHYAY B., HUMPHREY A. E. & TAGUCHI H. (1967) *Biotech. Bioeng.* **9**, 533

BARNES E. M. & INGRAM M. (1956) *J. appl. Bact.* **19**, 117

BARTHA R. & ORDAL E. J. (1965) *J. Bact.* **89**, 1015

BAUCHOP T. & ELSDEN S. R. (1960) *J. gen. Microbiol.* **23**, 457

BAUMBERGER J. P. (1939) *Cold Spring Harb. Symp.* **7**, 195

BEQUE W. J. & LICHSTEIN H. C. (1963) *Proc. Soc. exp. Biol. Med.* **114**, 625

BIRCH J. R. & PIRT S. J. (1969) *J. Cell Sci.* **5**, 135

BIRCH J. R. & PIRT S. J. (1970) *J. Cell Sci.* **7**, 661

BIRCH J. R. & PIRT S. J. (1971) *J. Cell Sci.* **8**, 693

BISSET K. A. (1950) *J. gen. Microbiol.* **4**, 413

BJERRUM J., SCHWARZENBACH G. & SILLEN L. G. (1958) *Stability Constants. Part I: Organic Ligands (Special Publication No. 6). Part II: Inorganic Ligands (Special Publication No. 7)*. Chemical Society, London

BLAKER G. J. (1971) *The Vitamin Nutrition of Mammalian Cells in Culture*. PhD Thesis, University of London

BLAKER G. J., BIRCH J. R. & PIRT S. J. (1971) *J. Cell Sci.* **9**, 529

BLAKER G. J. & PIRT S. J. (1971) *J. Cell Sci.* **8**, 701

BORICHEWSKI R. M. & UMBREIT W. W. (1966) *Arch. Biochem. Biophys.* **116**, 97

BORKOWSKI J. D. & JOHNSON M. J. (1967) *Appl. Microbiol.* **15**, 1483

BROOKS J. D. & MEERS J. L. (1973) *J. gen. Microbiol.* **77**, 513

BROMEL M. & TEODORO R. (1966) *Bact. Proc.* p. 78

BROWN C. M. & ROSE A. H. (1969) *J. Bact.* **99**, 371

BROWN C. M. & STANLEY S. O. (1972) *J. appl. Chem. Biotechnol.* **22**, 363

BRUNNER R., OBERZILL W. & MENZEL J. (1968) In *Continuous Cultivation of Micro-organisms* ed. Malek, I. *et al.*, p. 323. Academia, Prague

BRYSON V. (1952) *Science* **116**, 45

BUTTON D. K. (1969) *J. gen. Microbiol.* **58**, 15

BUTTON D. K. & GARVER J. C. (1966) *J. gen. Microbiol.* **45**, 195

CAMICI L., SERMONTI G. & CHAIN E. B. (1952) *Bull. Wld Hlth Org.* **6**, 265

CALCOTT P. H. & POSTGATE J. R. (1972) *J. gen. Microbiol.* **70**, 115

CALLOW D. S. & PIRT S. J. (1961) *J. appl. Bact.* **24**, 12

CAMPBELL L. L. & PACE B. (1968) *J. appl. Bact.* **31**, 24

CANALE R. P. (1970) *Biotech. Bioeng.* **12**, 353

CARLSSON J. (1971) *J. gen. Microbiol.* **67**, 69

CARTER B. L. A. & BULL A. T. (1969) *Biotechnol. Bioeng.* **11**, 785

CECCARINI C. & EAGLE H. (1961) *Proc. nat. Acad. Sci.* **68**, 229

CĚJKOVÁ A. (1965) *Folia Microbiologica* **10**, 246

CHAIN E. B. & GUALANDI G. (1954) *Rend. 1st. Super. Sanit.* (English edn) **17**, 5

CHAIN E. B., PALADINO S., CALLOW D. S., UGOLINI F. & VAN DER SLUIS J. (1952) *Bull. Wld Hlth Org.* **6**, 73

CHAO C. & REILLY P. J. (1972) *Biotech. Bioeng.* **14**, 75

CHESTER V. E. (1963) *Biochem. J.* **86**, 153

CHOUDHARY A. Q. & PIRT S. J. (1966) *J. gen. Microbiol.* **43**, 71

CHRISTIAN J. H. B. & HALL J. M. (1972) *J. gen. Microbiol.* **70**, 497

CHRISTIAN J. H. B. & SCOTT W. J. (1953) *Aust. J. Biol. Sci.* **6**, 565

CHRISTIAN J. H. B. & WALTHO J. A. (1964) *J. gen. Microbiol.* **35**, 205

CLARK W. M. (1960) *Oxidation–reduction Potentials of Organic Systems.* Williams & Wilkins, Baltimore

COHEN S. S. (1971) *Introduction to the Polyamines.* Prentice-Hall, New Jersey

CONTOIS D. E. (1959) *J. gen. Microbiol.* **21**, 40

COOPER A. L., DEAN A. C. R. & HINSHELWOOD C. (1968) *Proc. roy. Soc. B* **171**, 175

COOPER C. M., FERNSTROM G. A. & MILLER S. A. (1944). *Ind. Eng. Chem.* **36**, 504

COOPER K. E. (1963) In *Analytical Microbiology* ed. Kavanagh F., p. 46. Academic Press, New York

CURDS C. R. (1971) *Water Research* **5**, 793

CURDS C. R. & COCKBURN A. (1968) *J. gen. Microbiol.* **54**, 343

DAGLEY S. & HINSHELWOOD C. N. (1938) *J. Chem. Soc.* 1936–1942

DARLINGTON W. A. (1964) *Biotech. Bioeng.* **6**, 241

DAVIES H. C., KARUSH F. & RUDD J. H. (1965) *J. Bact.* **89**, 421

DAVISON M. J., DOWNIE J. A. & GARLAND P. B. (1972) *Biochem. J.* **129**, 46P

DAWES E. A. & RIBBONS D. W. (1964) *Bact. Rev.* **28**, 126

DAWES I. W., KAY D. & MANDELSTAM J. (1969) *J. gen. Microbiol.* **56**, 171

DAWES I. W. & THORNLEY J. H. M. (1970) *J. gen. Microbiol.* **62**, 49

DEAN A. C. R. (1969) In *Continuous Cultivation of Micro-organisms* ed. Malek I. *et al.*, p. 263. Academia, Prague

DEAN A. C. R. & HINSHELWOOD C. (1966) *Growth, Function and Regulation in Bacterial Cells.* Clarendon Press, Oxford

DEAN A. C. R. & MOSS D. A. (1970) *Chem.-Biol. Interactions* **2**, 281

DEAN A. C. R. & MOSS D. A. (1971) *Biochem. Pharmacol.* **20**, 1

DEAN A. C. R. & ROGERS P. L. (1967) *Biochim. biophys. Acta* **148**, 267

DE LEY J. (1962) In *Microbial Classification, 12th Symposium Soc. gen. Microbiol.* Cambridge University Press, Cambridge

DEMAIN A. L. (1972a) *Ann. Rev. Microbiol.* **26**, 369

DEMAIN A. L. (1972b) *J. appl. Chem. Biotechnol.* **22**, 345

DEMAIN A. L. & HENDLIN D. (1958) *J. Bact.* **75**, 46

DENBIGH K. G. (1965) *Chemical Reaction Theory, An Introduction.* Cambridge University Press, Cambridge

DE VRIES W., KAPTEIJN W. M. C., VAN DE BEEK E. G. & STOUTHAMER A. H. (1970) *J. gen. Microbiol.* **63**, 333

DICKS J. W. & TEMPEST D. W. (1967) *Biochim. biophys. Acta* **136**, 176

DIXON M. (1953) *Biochem. J.* **55**, 161

DIXON M. & ELLIOTT K. A. C. (1930) *Biochem. J.* **24**, 820

DIXON M. & WEBB E. C. (1967) *Enzymes*, 2nd edn. Longmans, London

DONALD C. (1952) *J. gen. Microbiol.* **7**, 211

DOWNIE J. A. & GARLAND P. (1972) *Biochem. J.* **129**, 47P

DREW S. W. & DEMAIN A. L. (1973) *Biotech. Bioeng.* **15**, 743

EDDY A. A., CARROLL T. C. N., DANBY C. J. & HINSHELWOOD C. (1951) *Proc. roy. Soc. B.* **138**, 219

EDDY A. A. & HINSHELWOOD C. (1951) *Proc. roy. Soc. B* **138**, 237

ELLWOOD D. C. (1971) *Biochem. J.* **121**, 349

ELLWOOD D. C. & TEMPEST D. W. (1972) *J. gen. Microbiol.* **73**, 395

ENOCH K. G. & LESTER R. L. (1972) *J. Bact.* **110**, 1032

EROSHIN V. K., HARWOOD J. H. & PIRT S. J. (1968) *J. appl. Bact.* **31**, 560

EVANS C. H. & BROWN E. G. (1973) *Proc. Biochem. Soc.* December, 1973

FAWCETT H. S. (1925) *Annals Appl. Biol.* **12**, 191

FENSOM A. H. & PIRT S. J. (1972) In *Fermentation Technology Today. Proc. IV I.F.S.*, p. 51. Society of Fermentation Technology, Japan

FEREN C. J. & SQUIRES R. W. (1969) *Biotech. Bioeng.* **11**, 583

FINN R. K. (1954) *Bacteriol. Rev.* **18**, 254

FINTER N. B. (1969) *J. gen. Virol.* **5**, 419

FLASCHKA H. A. (1964) *EDTA Titrations*, 2nd edn. Pergamon, Oxford

FREESE E., SHEU C. W. & GALLIERS E. (1973) *Nature* **241**, 321

FUJIO Y., SAMUICHI M. & UEDA S. (1973) *J. Ferm. Technol. (Japan)* **51**, 154

GALE E. F. & EPPS H. M. R. (1942) *Biochem. J.* **36**, 600

GALLUP D. M. & GERHARDT P. (1963) *Appl. Microbiol.* **11**, 506

GHOSH D. & PIRT S. J. (1954) *Rend. 1st Super. Sanit.* (English edn) **17**, 149

GILBY A. R. & FEW A. V. (1959) *J. gen. Microbiol.* **20**, 321

GLASSTONE S. & LEWIS D. (1964) *Elements of Physical Chemistry*, 2nd edn. Macmillan, London

GÖBEL F. & PFENNIG N. (1969) In *Continuous Cultivation of Micro-organisms* ed. Malek, I. et al., p. 337. Academia, Prague

GOMA G., PARAILLEUX A. & DURAND G. (1973) *J. Ferment. Technol. (Japan)* **51**, 616

GORDEE E. Z. & DAY L. E. (1972) *Antimicrobial Agents & Chemotherapy* **1**, 315

GORDON J., HOLMAN R. A. & McLEOD J. W. (1953) *J. Path. Bact.* **66**, 527

GOW J. S., LITTLEHAILES J. D., SMITH S. R. L. & WALTER F. B. (1973) *SCP Production from Methanol: Bacteria. International Symposium*, May 1973. Massachusetts Institute of Technology

GRIFFITHS J. B. & PIRT S. J. (1967) *Proc. roy. Soc. B* **168**, 421

HADJIPETROU L. P., GERRITS J. P., TEULINGS F. A. G. & STOUTHAMER A. H. (1964) *J. gen. Microbiol.* **36**, 139

HADJIPETROU L. P., GRAY-YOUNG T. & LILLY M. D. (1966) *J. gen. Microbiol.* **45**, 479

HADJIPETROU L. P. & STOUTHAMER A. H. (1965) *J. gen. Microbiol.* **38**, 29

HALL B. M. (1960) *Appl. Microbiol.* **8**, 378

HANSFORD G. S. & HUMPHREY A. E. (1966) *Biotech. Bioeng.* **8**, 85

HARRISON D. E. F. (1972) *Biochim. biophys. Acta*, **275**, 83

HARRISON D. E. F. (1973) *J. appl. Bact.* **36**, 301

HARRISON D. E. F. & LOVELESS J. E. (1971a) *J. gen. Microbiol.* **68**, 35

HARRISON D. E. F. & LOVELESS J. E. (1971b) *J. gen. Microbiol.* **68**, 45

HARRISON D. E. F., MACLENNAN D. G. & PIRT S. J. (1969) In *Fermentation Advances* ed. Perlman D. p. 117. Academic Press, New York

HARRISON D. E. F. & PIRT S. J. (1967) *J. gen. Microbiol.* **46**, 193

HARRISON D. E. F., TOPIWALA H. H. & HAMER G. (1972) *Fermentation Technology Today*, *Proc. IV I.F.S.*, p. 491. Society of Fermentation Technology, Japan

HARTE M. J. & WEBB F. C. (1967) *Biotech. Bioeng.* **9**, 205

HARWOOD J. H. (1970) *Studies on the Physiology of* Methylococcus capsulatus *growing on Methane*. PhD Thesis, University of London

HARWOOD J. H. & PIRT S. J. (1972) *J. appl. Bact.* **35**, 597

HATTORI K., YOKOO S. & IMADA O. (1972) *J. Ferment. Technol.* (*Japan*) **50**, 737

HERBERT D. (1958) In *Recent Progress in Microbiology, VII International Congress for Microbiology*, ed. Tunevall G., p. 381. Almquist & Wiksell, Stockholm

HERBERT D. (1961) In *Continuous Culture*, Monograph No. 12, p. 21. Society of Chemistry & Industry, London

HERBERT D. (1964) In *Continuous Cultivation of Micro-organisms* ed. Malek I. *et al.*, p. 23. Czechoslovak Academy of Sciences, Prague

HERBERT D. & PHIPPS P. J. (1974) *Proc. Soc. gen. Microbiol.* **1/3**

HERNANDEZ E. & JOHNSON M. J. (1967) *J. Bact.* **94**, 996

HERNANDEZ E. & PIRT S. J. (1975) *J. appl. Chem. Biotechnol.* in press

HILL A. V. (1928) *Proc. Roy. Soc. B* **104**, 39

HILLS G. M. & SPURR E. D. (1952) *J. gen. Microbiol.* **6**, 64

HINSHELWOOD C. N. (1946) *The Chemical Kinetics of the Bacterial Cell*. Clarendon Press, Oxford

HINSHELWOOD C. N. (1952) *J. chem. Soc.* **745**

HINSHELWOOD C. N. (1953) *J. chem. Soc.* **1947**

HOBSON P. N. (1969) In *Methods in Microbiology 3B*, ed. Norris J. R. & Ribbons D. W., p. 133. Academic Press, London

HOBSON P. N. & MANN S. O. (1970) In *Automation, mechanization and data handling in microbiology*. Society for Applied Bacteriology, Academic Press, London

HOLZER H. & DÜNTZE W. (1971) *Ann. Rev. Biochem.* **40**, 345

HOSPODKA J. (1966) *Biotech. Bioeng.* **8**, 117

HOŠŤÁLEK Z., TINTEROVA M., JECHOVÁ V., BLUMAUEROVÁ M., SUCHÝ J. & VANĚK Z. (1969) *Biotech. Bioeng.* **11**, 539

HOTTA K. & TAKAO S. (1973) *J. Ferm. Technol.* (*Japan*) **51**, 12

HROMATKA O. (1952) *Chemiker Ztg* **76**, 776

HSIEH D. P. H., SILVER R. S. & MATELES R. I. (1969) *Biotech. Bioeng.* **11**, 1

HUMPHREY A. E. & ERICKSON L. E. (1972) *J. appl. Chem. Biotechnol.* **22**, 125

HUNGATE R. E. (1969) In *Methods in Microbiology 3B*, ed. Norris J. R. & Ribbons D. W., p. 117. Academic Press, London

HUNTER K. & ROSE A. H. (1972) *J. appl. Chem. Biotechnol.* **22**, 527

HUTNER S. H. (1972) *Ann. Rev. Microbiol.* **26**, 313

INGRAM M. (1947) *Proc. roy Soc. B* **134**, 181

INGRAHAM J. L. (1958) *J. Bact.* **76**, 75

ISHIZAKI A., SHIBAI H., HIROSE Y. & SHIRO T. (1973) *Agric. Biol. Chem.* **37**, 99

JACOB H. E. (1970) In *Methods in Microbiology 2*, p. 91. Academic Press, London

JENSEN A. L., DARKEN M. A., SCHULTZ J. S. & SHAY A. J. (1963) *Antimicrobial Agents and Chemotherapy* 49

JOHNSON M. J. (1964) *Chem. & Ind.* 1532

JOHNSON M. J. (1967a) *Science* **155**, 1515

JOHNSON M. J. (1967b) *J. Bact.* **94**, 101

JONES G. K., JANSEN F. & McKAY A. J. (1973) *J. gen. Microbiol.* **74**, 139

JOST J. L., DRAKE J. F., FREDRICKSON A. G. & TSUCHIYA H. M. (1973) *J. Bact.* **113**, 834

KANTOROWICZ O. (1951) *J. gen. Microbiol.* **5**, 276

KELLY D. P. (1967) *Sci. Prog. (Oxford)* **55**, 35

KEPES A. & COHEN G. N. (1962) In *The Bacteria. IV. The Physiology of Growth*, p. 179. Academic Press, New York

KHANNA R., TEWARI K. K. & KRISHNAN P. S. (1963) *Archiv. Mikrobiol.* **44**, 352

KIHARA H., KLATT O. A. & SNELL E. E. (1952) *J. biol. Chem.* **197**, 801

KIHARA H. & SNELL E. E. (1952) *J. biol. Chem.* **197**, 791

KINOSHITA S. (1972) In *The Microbial Production of Amino Acids* ed. Yamada K. *et al.* John Wiley, New York

KITAI A., TONE H. & OZAKI A. (1969) *Biotech. Bioeng.* **11**, 911

KLEIN F., MAHLANDT B. G. & LINCOLN R. E. (1971) *Appl. Microbiol.* **22**, 145

KLOTZ I. M. & GUTMANN H. R. (1945) *J. Amer. chem. Soc.* **67**, 558

KLUYVER A. J. & PERQUIN L. H. C. (1933) *Biochemische Zeit.* **266**, 68

KNIGHT B. C. J. G. & FILDES P. (1930) *Biochem. J.* **24**, 1496

KOBAYASHI K., IKEDA S., HIROSE Y. & KINOSHITA K. (1967) *Agric. Biol. Chem.* **31**, 1448

KOCH A. L. & COFFMAN R. (1970) *Biotech. Bioeng.* **12**, 651

KORMANČÍKOVÁ V., KOVÁČ L. & VIDOVÁ M. (1969) *Biochim. biophys. Acta.* **180**, 9

KOSER S. A. (1968) *Vitamin Requirements of Bacteria and Yeasts.* Charles C. Thomas, Springfield, Ill.

KRUSE P. F. & MIEDEMA E. (1965) *J. Cell Biol.* **27**, 273

KUROWSKI W. M. (1974) *Transformation of Sucrose to 3-Ketosucrose by* Agrobacterium tumefaciens *and Stability of the Glucoside-3-dehydrogenase Activity in Non-growing Cells.* PhD Thesis, University of London

KUROWSKI W. M., FENSOM A. H. & PIRT S. J. (1973) *J. gen. Microbiol.* **75/2**, xv

KUROWSKI W. M. & PIRT S. J. (1971) *J. gen. Microbiol.* **68**, 65

LANE A. G. & PIRT S. J. (1973) *J. appl. Chem. Biotechnol.* **23**, 309

LARSEN H. (1967) *Advances in Microbial Physiology*, ed. Rose A. H. & Wilkinson J. F., **1**, 97. Academic Press, London

LAYNE E. (1957) In *Methods in Enzymology* ed. Colowick S. P. & Kaplan N. O., p. 447. Academic Press, New York

LICHSTEIN H. C. (1960) *Ann. Rev. Microbiol.* **14**, 17

LIGHT A. (1972) *J. appl. Chem. Biotechnol.* **22**, 509

LODGE R. M. & HINSHELWOOD C. N. (1943) *J. chem. Soc.* 213

LOWRY H. O., ROSENBROUGH N. J., FARR A. C. & RANDALL R. J. (1951) *J. biol. Chem.* **193**, 265

LUEDEKING R. & PIRET E. L. (1959) *J. biochem. microbiol. Technol. Engng* **1**, 393

LUNGE G. & KEANE C. A. (1908) *Technical Methods of Chemical Analysis* **1/1**, p. 782. Gurney & Jackson, London

LWOFF A. & MONOD J. (1947) *Ann. Inst. Past.* **73**, 323

MAALOE O. & KJELDGAARD N. O. (1966) *Control of Macromolecular Synthesis.* Benjamin W. A., New York

MACKINTOSH I. P. (1973) *Mathematical and Mechanical Models simulating Conditions of Bacterial Growth in the Human Bladder.* PhD Thesis, University of London

MACLENNAN D. G., OUSBY J. C., VASEY R. B. & COTTON N. T. (1971) *J. gen. Microbiol.* **69**, 395

MACLENNAN D. G. & PIRT S. J. (1966) *J. gen. Microbiol.* **45**, 289

MAGER J., KUCZNSKI M., SCHUTZBERG G. & AVI-DOR Y. (1956) *J. gen. Microbiol.* **14**, 69

MANDELSTAM J. (1958) *Biochem. J.* **69**, 103

MANDELSTAM J. (1960) *Bacteriol. Rev.* **24**, 289

MANFREDINI R. & WANG D. I. C. (1972) *Biotech. Bioeng.* **14**, 267

MARGALITH P. & PAGANI H. (1961) *Appl. Microbiol.* **9**, 325

MARKHAM E. & BYRNE W. J. (1969) In *Continuous Cultivation of Micro-organisms* ed. Malek I. *et al.*, p. 117. Academia, Prague

MARR A. G. & INGRAHAM J. L. (1962) *J. Bact.* **84**, 1260

MARR A. G., NILSON E. H. & CLARK D. J. (1963) *Ann. N.Y. Acad. Sci.* **102**, Art. 3, 536

MATELES R. I., RYU D. Y. & YASUDA T. (1965) *Nature* **208**, 263

MAUCK J., CHAN L. & GLASER L. (1971) *J. biol. Chem.* **246**, 1820

MEERS J. L. (1971) *J. gen. Microbiol.* **67**, 359

MEERS J. L. & BROOKS J. D. (1973) *J. gen. Microbiol.* **75**, viii

MEERS J. L. & TEMPEST D. W. (1968) *J. gen. Microbiol.* **52**, 309

MEYNELL G. G. & MEYNELL E. (1965) *Theory and Practice in Experimental Bacteriology.* Cambridge University Press, Cambridge

MILES R. J. & PIRT S. J. (1969) *Biochem. J.* **114**, 10P

MILES R. J. & PIRT S. J. (1973) *J. gen. Microbiol.* **76**, 305

MIMURA A., KAWANO T. & KODAIRA R. (1969) *J. Ferm. Technol.* **47**, 229

MITCHELL P. (1973) *J. Bioenergetics* **4**, 63

MITCHELL P. & MOYLE J. (1956) In *Bacterial Anatomy, 6th Symposium Soc. gen. Microbiol.*, p. 150. Cambridge University Press, Cambridge

MONOD J. (1942) *Recherches sur la Croissance des Cultures Bactériennes.* 2nd edn. Hermann, Paris

MONOD J. (1950) *Ann. Inst. Past.* **79**, 390

MOSES V. & SYRETT P. J. (1955) *J. Bact.* **70**, 201

MOSS F. J., RICKARD P. A. D., BEECH G. A. & BUSH F. E. (1969) *Biotech. Bioeng.* **11**, 561

MOYER A. J. (1953) *Appl. Microbiol.* **1**, 1

MUNK V., DOSTALEK M. & VOLFOVÁ O. (1973) *Biotech. Bioeng.* **11**, 383

NAGAI S. & AIBA S. (1972) *J. gen. Microbiol.* **73**, 531

NG F. M. W. & DAWES E. A. (1973) *Biochem. J.* **132**

NG H., INGRAHAM J. L. & MARR A. G. (1962) *J. Bact.* **84**, 331

NOGUCHI Y. & JOHNSON M. J. (1961) *J. Bact.* **82**, 538

NOVICK A. (1958) In *Continuous Cultivation of Micro-organisms*, Symposium, p. 29. Czechoslovak Academy of Sciences, Prague

NOVICK A. & SZILARD L. (1950) *Science*, **112**, 715

O'BRIEN R. W. & STERN J. R. (1969) *J. Bact.* **98**, 388

O'SULLIVAN C. Y. & PIRT S. J. (1973) *J. gen. Microbiol.* **76**, 65

OKUMURA S. (1972) In *The Microbial Production of Amino Acids*, ed. Yamada K. *et al.* John Wiley, New York

OWEN S. P. & JOHNSON M. J. (1955) *Appl. Microbiol.* **3**, 375

PALADINO S. (1954) *Rend. 1st Super. Sanit.* (English edn) **17**, 145

PALUMBO S. A., JOHNSON M. G., RIECK V. T. & WITTER L. D. (1971) *J. gen. Microbiol.* **66**, 137

PASTEUR L. (1869) *Annales de Chimie et de Physique, 3ᵉ Série*, **58**, 324

PAUL J. (1965) *Cell and Tissue Culture.* 3rd edn. Livingstone, Edinburgh

PERLMAN D. (1973) *Process Biochem.* **8**/7, 18

PETERS V. J., PRESCOTT J. M. & SNELL E. E. (1953) *J. biol. Chem.* **202**, 521

PHELPS A. (1936) *J. exp. Zool.* **72**, 479

PHILLIPS D. H. (1966) *Biotech. Bioeng.* **8**, 456

PHILLIPS D. H. & JOHNSON M. J. (1959) *Ind. Eng. Chem.* **51**, 83

PHILLIPS D. H. & JOHNSON M. J. (1961) *J. biochem. microbiol. Technol. & Engng* 3, 261

PIPER W. O. (1962) *Appl. Microbiol.* 10, 281

PIRT S. J. (1957) *J. gen. Microbiol.* 16, 59

PIRT S. J. (1965) *Proc. roy. Soc. B* 163, 224

PIRT S. J. (1966) *Proc. roy. Soc. B* 166, 369

PIRT S. J. (1967) *J. gen. Microbiol.* 47, 181

PIRT S. J. (1969) In *Microbial Growth, 19th Symposium Soc. gen. Microbiol.* Cambridge University Press, Cambridge

PIRT S. J. (1971) *Biochem. J.* 121, 293

PIRT S. J. (1972a) *Year Book 1972*, p. 119. Association of River Authorities. London

PIRT S. J. (1972b) *J. appl. Chem. Biotechnol.* 22, 55

PIRT S. J. (1973a) *J. appl. Chem. Biotechnol.* 23, 389

PIRT S. J. (1973b) *J. gen. Microbiol.* 75, 245

PIRT S. J. (1974) *J. appl. Chem. Biotechnol.* 24, 415

PIRT S. J. & BAZIN M. J. (1972) *Nature* 239, 290

PIRT S. J. & CALLOW D. S. (1958a) *J. appl. Bact.* 21, 188

PIRT S. J. & CALLOW D. S. (1958b) *J. appl. Bact.* 21, 211

PIRT S. J. & CALLOW D. S. (1959) *Nature* 184, 307

PIRT S. J. & CALLOW D. S. (1960) *J. appl. Bact.* 23, 87

PIRT S. J., CALLOW D. S. & GILLETT W. A. (1957) *Chem. & Ind.* 730

PIRT S. J. & KUROWSKI W. M. (1970) *J. gen. Microbiol.* 63, 357

PIRT S. J. & RIGHELATO R. C. (1967) *Appl. Microbiol.* 15, 1284

PIRT S. J. & THACKERAY E. J. (1964) *Exp. Cell Res.* 33, 396

PIRT S. J., THACKERAY E. J. & HARRIS-SMITH R. (1961) *J. gen. Microbiol.* 25, 119

POSTGATE J. R. & HUNTER J. R. (1962) *J. gen. Microbiol.* 29, 233

POSTGATE J. R. & HUNTER J. R. (1964) *J. gen. Microbiol.* 34, 459

POWELL E. O. (1956) *J. gen. Microbiol.* 15, 492

POWELL E. O. (1963) *J. Sci. Food Agric.* 1

POWELL O. & LOWE J. R. (1964) In *Continuous Cultivation of Micro-organisms*, ed. Malek I. *et al.* p. 45. Czechoslovak Academy of Sciences, Prague

PRESCOTT J. M., PETERS V. J. & SNELL E. E. (1953) *J. biol. Chem.* 202, 533

RATLEDGE C. (1971) *Biochem. biophys. Res. Commun.* 45, 856

RATLEDGE C. & CHAUDHRY M. A. (1971) *J. gen. Microbiol.* 66, 71

RAULIN M. J. (1869) *Annales des Sciences Naturelles, 5ᵉ Série*, 11, 93

REESE E. T. (1972) In *Enzyme Engineering*, ed. Wingard L. B. Interscience, New York

ŘIČICA J. (1958) In *Continuous Cultivation of Micro-organisms, Symposium*, p. 75. Czechoslovak Academy of Sciences, Prague

RIGHELATO R. C., TRINCI A. P. J., PIRT S. J. & PEAT A. (1968) *J. gen. Microbiol.* 50, 399

RIGHELATO R. C. & VAN HEMERT P. (1969a) In *Continuous Cultivation of Micro-organisms*, ed. Malek I. *et al.*, p. 437. Academia, Prague

RIGHELATO R. C. & VAN HEMERT P. A. (1969b) *J. gen. Microbiol.* 58, 403

ROBERTS R. B., ABELSON P. H., COWIE D. B., BOLTON E. T. & BRITTEN R. J. (1955) *Studies of Biosynthesis in* Escherichia coli. Carnegie Institute Publ. 607, Washington

ROBINSON R. A. & STOKES R. H. (1959) *Electrolyte Solutions*. Butterworths, London

ROGOSA M., FRANKLIN J. G. & PERRY K. D. (1961) *J. gen. Microbiol.* 25, 473

ROSENBERGER R. F. & ELSDEN S. R. (1960) *J. gen. Microbiol.* 22, 726

ROUGHTON F. J. W. & SCHOLANDER P. F. (1943) *J. Biol. Chem.* 148, 541

ROWLEY B. I. & PIRT S. J. (1972) *J. gen. Microbiol.* 72, 553

ROYSTON M. G. (1966) *Process Biochem.* 1, 215

RUBIN M. (1961) *Fed. Proc.* 20 (Suppl. I), 149

SACKS L. E. & PENCE J. W. (1958) *J. gen. Microbiol.* **19,** 542

SANFORD K. K., EARLE W. R., EVANS V. J., WALTZ H. K. & SHANNON J. E. (1951) *J. nat. Cancer. Inst.* **11,** 773

SATO M., NAKAHARA T. & YAMADA K. (1972) *Agric. Biol. Chem.* **36,** 2025

SCOTT W. J. (1953) *Aust. J. Biol. Sci.* **6,** 549

SCOTT W. J. (1957) *Advances in Food Research* **7,** 83

SCHULDINER S., PIERSMA B. J. & WARNER T. B. (1966) *J. electrochem. Soc.* **113,** 573

SCHULTZ J. S. (1964) *Appl. Microbiol.* **12,** 305

SCHULTZ J. S. & GADEN E. L. (1956) *Ind. Eng. Chem.* **48,** 2209

SCHULTZ J. S. & GERHARDT P. (1969) *Bact. Rev.* **33,** 1

SEELEY H. W. & VANDEMARK P. J. (1951) *J. Bact.* **61,** 27

SHEHATA T. E. & MARR A. G. (1971) *J. Bact.* **107,** 210

SHINDELA A., BUNGAY H. R., KRIEG N. R. & CULBERT K. (1965) *J. Bact.* **89,** 693

SILLÉN L. G. & MARTELL A. E. (1971) *Stability Constants of Metal Ion Complexes,* Supplement No. 1 (Special Publication No. 25). Chemical Society, London

SISTROM W. R. (1960) *J. gen. Microbiol.* **22,** 778

SLATOR A. (1921) *J. chem. Soc.* **119,** 115

SMITH C. G. & JOHNSON M. J. (1954) *J. Bact.* **68,** 346

SNELL E. E. (1949) *Ann. Rev. Microbiol.* **3,** 97

SOMEYA J., MURAKAMI T., TAGAYA N., FUTAI N. & SONODA Y. (1970) *J. Ferm. Technol. (Japan)* **48,** 291

STANDING C. N., FREDRICKSON A. G. & TSUCHIYA H. M. (1972) *Appl. Microbiol.* **23,** 354

STANIER R. Y., DOUDOROFF M. & ADELBERG E. A. (1963) *General Microbiology* 2nd edn. Macmillan, London

STEINBERG R. A. (1956) *Nat. Acad. Sci.–Nat. Res. Council Publ.* **514,** 1

STICKLAND L. H. (1951) *J. gen. Microbiol.* **5,** 698

STOUTHAMER A. H. (1973) *Ant. v. Leeuwenhoek* **39,** 545

STOUTHAMER A. H. & BETTENHAUSEN C. (1973) *Biochim. biophys. Acta.* **301,** 53

STRANGE R. E. & DARK F. A. (1965) *J. gen. Microbiol.* **39,** 215

STRANGE R. E., DARK F. A. & NESS A. G. (1961) *J. gen. Microbiol.* **25,** 61

SUZUKI T., TANAKA K. & OKUMURA S. (1969) *Production of Sugars and Amino Acids from Hydrocarbons and Petrochemicals by Micro-organisms.* Abstract of World Petroleum Congress, MOSCOW 1971

TAJIMA K. & YOSHIZUMI H. (1972) *J. Ferment. Technol. (Japan)* **50,** 764

TAKAHASHI J., UEMURA N. & UEDA K. (1970) *Agric. biol. Chem.* **34,** 32

TEMPEST D. W. (1969) In *Microbial Growth, 19th Symposium Soc. gen. Microbiol.* p. 87. Cambridge University Press, Cambridge

TEMPEST D. W. & HERBERT D. (1965) *J. gen. Microbiol.* **39,** 355

TEMPEST D. W., HERBERT D. & PHIPPS P. J. (1967) In *Microbial Physiology and Continuous Culture* ed. Powell E. O. *et al.,* p. 240. H.M.S.O., London

TEMPEST D. W. & MEERS J. L. (1968) *J. gen. Microbiol.* **54,** 319

THOMAS T. D. & BATT R. D. (1968) *J. gen. Microbiol.* **50,** 367

THURSTON C. F. (1972) *Process Biochem.* **7/8,** 18

TJÖTTA E. (1966) *Acta Path. microbiol. scand.* **68,** 451

TOPIWALA H. H. & HAMER G. (1971) *Biotech. Bioeng.* **13,** 919

TOPIWALA H. & SINCLAIR C. G. (1971) *Biotech. Bioeng.* **13,** 795

TOVEY M. G., MATHISON G. E. & PIRT S. J. (1973) *J. gen. Virol.* **20,** 29

TOWNSLEY P. M. & NEILANDS J. B. (1957) *J. biol. Chem.* **224,** 695

TRINCI A. P. J. (1969) *J. gen. Microbiol.* **57,** 11

TRINCI A. P. J. (1970) *Arch. Mikrobiol.* **73,** 353

TRINCI A. P. J. (1971) *J. gen. Microbiol.* **67,** 325

TRINCI A. P. J. (1973) *Arch. Mikrobiol.* **91**, 113

TRINCI A. P. J. & RIGHELATO R. C. (1970) *J. gen. Microbiol.* **60**, 239

TRUESDALE G. A., DOWNING A. L. & LOWDEN G. F. (1955) *J. appl. Chem.* **5**, 53

TSAO G. T. (1970) *Biotech. Bioeng.* **12**, 51

TSUCHIYA H. M., DRAKE J. F., JOST J. L. & FREDRICKSON A. G. (1972) *J. Bact.* **110**, 1147

TYRELL E. A., MACDONALD K. E. & GERHARDT P. (1958) *J. Bact.* **75**, 1

UEMURA N., TAKAHASHI J. & UEDA K. (1969) *J. Ferm. Technol. (Japan)* **47**, 220

UGOLINI F., UGOLINI G. & CHAIN E. B. (1959) *Selected Scientific Papers, 1st Super. Sanit.* **II**, Part I, 1

ULRICH K. & MOORE G. E. (1965) *Biotech. Bioeng.* **7**, 507

VAN UDEN N. (1967) *Arch. Mikrobiol.* **58**, 155

WALLACE A. (1962) In *A Decade of Synthetic Chelating Agents in Inorganic Plant Nutrition*, ed. Wallace A. Geigy Agricultural Chemicals, Ardsley, N.Y.

WARE G. C. (1951) *J. gen. Microbiol.* **5**, 880

WATANABE I. & OKADA S. (1967) *J. cell. Biology* **32**, 309

WATSON T. G. (1970) *J. gen. Microbiol.* **64**, 91

WATSON T. G. (1972) *J. appl. Chem. Biotechnol.* **22**, 229

WAYMOUTH C. (1956) *J. nat. Cancer Inst.* **17**, 305

WILDIERS E. (1901) *Cellule* **18**, 311

WILKINSON T. G. & HARRISON D. E. F. (1973) *J. appl. Bact.* **36**, 309

WILKINSON J. F. & MUNRO A. C. S. (1967) In *Microbial Physiology and Continuous Culture*, ed. Powell E. O. *et al.*, p. 173. H.M.S.O., London

WILLIS A. T. (1969) *Methods in Microbiology 3B*, ed. Norris J. R. & Ribbons D. W., p. 80. Academic Press, London

WILSON D. O. & REISENAUER H. M. (1970) *J. Bact.* **102**, 729

WIMPENNY J. W. T. (1969) in *Microbial Growth 19th Symposium Soc. gen. Microbiol.*, p. 161. Cambridge University Press, Cambridge

WISEMAN G. M., VIOLAGO F. C., ROBERTS E. & PENN I. (1966) *Canad. J. Microb.* **12**, 521

WODZINSKI R. J. & FRAZIER W. C. (1960) *J. Bact.* **79**, 572

WODZINSKI R. J. & FRAZIER W. C. (1961) *J. Bact.* **81**, 353

WRIGHT D. G. & CALAM C. T. (1968) *Chem. & Ind.* 1274

YASUDA S., SATOH K., ISHII T. & FURUYA T. (1972) *Fermentation Technology Today*, p. 697. Society of Fermentation Technology, Japan

YEOH H. T., BUNGAY H. R. & KRIEG N. R. (1968) *Canad. J. Microbiol.* **14**, 491

YOSHIDA F., YAMANE T. & NAKAMOTO K. (1973) *Biotech. Bioeng.* **15**, 257

Index

Absorptiometry 19
Acetate 74
Acetic acid bacteria 111
Activated sludge process (*see also* Effluent purification) 28
Activation energy 137–140
Activator of metabolism 170, 184
  product 202
Activity coefficient 84
Aeration methods (*see also* Oxygen) 94–106
Aerobe 113
Affinity for substrate 9
Agitation (*see also* Aeration methods) 95
Air lift 96
Air sparging 98
Alkanes (*see* Hydrocarbons)
Amino acids 120
  pool in cell 119
Anaerobic growth 114–116
Anaerobic methods 114
Antifoam agents 100, 102
Antigens 142
ATP 64, 112
  yield 73–74, 82
Autolysis 62, 187
Autosynthesis model 170, 218, 243–249
  and environmental change 246–248
  network theorem 248
Autotroph 63, 74, 120
Azotobacter 69, 70, 71

Baffled aeration 95, 103
Batch culture 22–25, 35, 57, 60, 72, 92, 159, 171, 173, 176, 199, 211, 243
  mathematical model 24
Brewing 50
Biomass
  concentration 47, 221
  definition 15
  effect on aeration 101, 104
  estimation 15–21
  feedback 26, 27, 42, 45–50, 55, 164

  maximum 7, 13
  output rate 34–37
  packing density 13–14
  production in packed column 228
  terminology 15
Bios factors 2, 117
Biphasic culture 221
Borate 127
Butanediol production 111

Carbon dioxide 43
  carbonate equilibria 76–77
  pressure effects 77–78, 242
  saturation constant 77
  solubility 76
  sparing agents 77
  supply 75–76
Carbon sources 63–80
  assimilation and dissimilation 64–66
  growth yield 64
  mixed 75, 168
  saturation constant 12
Catabolite repression 167
Catalase 112
Cell walls 152
Chelation of metal ions 128–133
Chemiosmotic theory 143
Chemolithotroph 63
Chemoorganotroph 63
Chemostat 29–56, 42, 58–60, 62, 91, 108, 145, 162, 172, 174, 176, 178, 183, 189, 199, 207, 208, 211, 243
  apparatus effect 39
  with biomass feedback 45–50
  deviations from theory 38
  inhibition in 172–184
  purposes 41
  in series 50–56
Choline 7, 8
Closed system 22
Cobalt 127
Colony growth 234–242
  area growth law 237

Colony growth (*contd*)
 of bacteria 236–239
  effect of agar depth 237
  effect of oxygen 238
  effect of specific growth rate 238
  effect of substrate concentration 237, 241
 of fungi 239–241
 linear growth law 236, 239
 model 234–236
 perimeter instability 239
 threshold substrate concentration 237
Commensalism 199
Competition between species (*see* Mixed cultures)
Competitive inhibition (*see* Inhibition)
Composition of microbes 142, 146, 152
Conditioned medium 198
Contact inhibition of growth 13
Contact time of gas 95
Copper 127
Counting cells 20
Critical dilution rate (*see* Dilution rate)
Cube root growth law 230
Cytochromes 110

Death of cells 13, 57–62
 in chemostat 58
Death rate 57, 140
 specific 57
Decline phase 23
Defined media (*see* Medium)
Degree of multiplication 5
Dialysis culture 218–222
Diauxie 25
Differentiation into spores 57
Diffusion capsule 221–222
Diffusion count of cells 21
Dilution rate 27, 32
 critical 28, 33, 47, 49, 51, 55
 overall 51, 52, 55
 partial 51
Dissolved oxygen concentration (*see* Oxygen)
Dissolved oxygen tension (*see* Oxygen)
Dormant cells 57, 189
Doubling time of biomass 5
 apparent 58
 reciprocal 6
Draught tube 96

Economic coefficient 7
EDTA 129
Effluent purification 28, 208
E_h (*see* Redox potential)

Electron acceptors 63
Electron donors 63
Endogenous metabolism (*see* Metabolism)
Energy source 63–80
 effect on product formation 167
 growth yield 64
 metabolic fate 74
 reserve (*see* Storage product)
 saturation constant 12, 64
Error in cell counts 20
Exponential growth 5–6
Extinction 19

Facultative anaerobes 111, 113
Fed batch culture 211–218
 repeated 216
Fermentation 64
Ferricyanide 112
Film of biomass 223–230
 theoretical 224–226
Foam, breaking 101–102
 effects of 101
 formation 101

Galactosidase-β 140, 189
Gas absorption (*see* Oxygen transfer)
Gas diffusion through cotton 104
Gas transfer (*see* Oxygen transfer)
Glucose oxidase 90
Gram stain 186
Growth
 curve 23, 24
 factor 117, 118, 141
 lag (*see* Lag period)
 limiting substrate 13, 32, 54
 parameters 4
 phases 22–23
 rate (*see* Specific growth rate)
 single point estimate 23
 unit of fungal hypha 240
Growth yield (*see also individual substrates*) 6–7, 12–13, 18, 39, 64, 118, 122
 true 66

Haematocrit 17
Henry's constant (H) 83
 for carbon dioxide 76
 law 83
 for oxygen 83
Heterotroph 59
History 1
Hofmeister series 153
Hold up of gas 95

Hormone 117, 123
Hydrocarbons 20, 78–80, 100, 217
    dispersion 80, 103
    emulsification 80
    and oxygen demand 82
    solubility 78
Hydrogen acceptor (*see also* Electron
    acceptor)
    electrode 85
Hydrogen ion concentration 143–146
    (*see also* pH)
Hyperbaric oxygen pressure 109

Impeller design 97
Incremental feed 81
Inhibition 170–184
    affecting growth yield 180
    competitive 120, 170, 171, 176
    by ethanol 175
    non-competitive 120, 170, 173, 178
    by product 13, 50, 176–180, 202, 206,
        221
    by substrate 45, 181–184
    by trace metals 125
Inoculum effects 115, 118, 196, 201
Inosine 78
Insulin 123, 185
Interdependent syntheses 243
Interstitial volume of packed cells 16
Intestinal flora 28
Involution forms 186
Iodide in nutrition 127
Ionic strength 69, 147, 153
Iron nutrition 125–127, 133

Ketoglutarate 74
Kerosene 101
$K_L a$ value 88–93, 96
    biomass effect 101
    foam effect 100
    measurement 90–93
    power requirement 101
    surfactant effect 100
    temperature effect 100
    viscosity effect 100
K_s value (*see* Saturation constant)

Lag period before growth 11–13, 23,
    194–198
    apparent 194
    definition 11
    early 196
    inoculum effects 196
    late 196
    metabolism during 197

effect of temperature 197
    true 194
Light energy 63
Light scattering 19
Liquid film resistance (*see* Oxygen transfer
    rate)
Logarithmic growth (*see* Exponential
    growth)
Low growth rate effects 186–193
Lysis 16

'M' concentration 13
Macro-regulation of metabolism 250
Magnesium 124, 201
Maintenance energy 39, 64, 66–74, 110,
    188, 189, 213
    effect of temperature on 141
    effect on Y_{ATP} 73
Maintenance rate (*see* Specific mainten-
    ance rate)
Major elements 117
Manganese nutrition 125
Manometric technique 81
Mass transfer (*see* Oxygen transfer)
Master reaction 44
Medium (*see also* Minimal medium)
    defined 1, 117, 118, 253
    stability 133
    undefined 117, 135
Metabolic quotient (q) 7, 18
    for energy source 67
    for oxygen (*see* Respiration rate)
Metabolism in non-growing cells 187
    endogenous 68, 187
Metal ion buffers (*see also* Chelation of
    metal ions) 132
Methylcellulose 133
Methylene blue 86
Michaelis–Menten constant 8
Microaerophilic 113
Microbiological assay 118–119
Microbe 1
Microregulation of metabolism 250
Minimal medium 21, 118, 133, 195
Minimum growth rate (*see* Specific growth
    rate)
Minor elements 117
Mixed carbon sources (*see* Carbon
    sources)
Mixed cultures 199–210
    competition for substrate 200
    controlled competition 201
    crossover point 200
    diversity of population 209
    non-competitive growth 203
    product-limited 203–205

Mixing 39, 95
Molybdenum 127
Monod relation 9, 241
  deviation from 11
Monolayer culture 223
Morphology 146
Mutation rate 208

Natural medium 117
Nephelometry 19
Network theorem (see Autosynthesis model)
Nickel 127
Nitrate 112
Nitrogen
  fixing 69
  limited growth 121
  sources 119, 168
Non-growing cells 110
Nutrient deficiency (see also Substrate-limited growth) 168
Nutrient groups 117
Nutrition 117–136
  effect of temperature on 141

Opacity 19
Open system 22, 25
Optical density 19
Oscillations in populations 124, 207
Osmotic coefficient 148–149
Osmotic pressure and water activity (see also Tonicity) 148–149, 153–155
Osmotic swelling 20
Oxidase 81, 83, 110
Oxidation-reduction (see also Redox potential) 64
Oxygen, absorption rate (see also Oxygen transfer rate)
  activity 84
  concentration 83
  critical concentration 109, 111
  demand 81–83
  determination 84–85
  dissolved oxygen concentration 84
  dissolved oxygen tension (DOT) 84
  effects on metabolism 110–112
  effect on viability 111
  E_h value 86–87
  electrode 81, 84–85
  growth yield 64, 65
  inhibitory effects (see also Oxygen toxicity) 112–114
  limited growth 93, 107, 229, 238
  redox potential 86
  saturation constant 109

  solubility 83–84
  solution rate (see Oxygen transfer rate)
  substitutes 112
  supply (see Oxygen transfer)
  tension 84
  toxicity 238
Oxygenase 81, 83
Oxygen transfer (see also K_La value and Aeration methods) 82, 87–93
  gas to liquid 87
  liquid to biomass 89
  liquid film resistance 89, 90, 91, 95
  power requirement 101
  rate measurement 90–93
  stationary liquid film theory 88
  two film theory 89

Packed column 224–230
  long formula 228
  short formula 228
Parasitism 199
Pasteur effect 107
Pellet growth 230–233
  causes 232
Penicillin 94, 141, 158, 159, 166, 168, 218
Peptides 120
pH (see also Hydrogen ion concentration) 44
  control 143–144
  effects on cultures 144–146, 250
  effect on metal chelation 129
  gradient 144
Phenol 184
Phosphorus 123
Photolithotroph 63
Photorganotroph 63
Plug-flow culture 25–28, 29, 55
P/O ratio 74, 82
Poisson distribution 20
Polypropylene glycol 102
Population growth rate 4
Potassium 123
Power ratio 101
Power requirement (see Oxygen transfer)
Precursors 168
Predator–prey interactions 205–208
Prey (see Predator–prey interactions)
Product classification 157
Product decomposition 159
Product formation 156–169, 203, 214
  in batch culture 159
  in chemostat culture 162–166
  concentration by dialysis 220
  decay of 160, 164–167
  environmental effects on 167–169
  in fed batch culture 214–217

Product formation (*contd*)
  growth-linked 157, 176, 178
  non-growth-linked 158, 176
  and temperature 141
  transient 162
Product inhibition (*see* Inhibition)
Product yield 158, 164, 176, 178
Protist 1
Protein turnover 70, 167, 187
Proton, conductor 146
  motive force 143
Protozoa 205
Psychrophile 140
Pyrogallol 90
Pyruvate 74

q_{O2} (*see* Respiration rate)
Quasi-steady state 211
Quick vinegar process 224

Raoult's law 148
Redox potential (E_h) 81, 85–87
  for anaerobic growth 114
  indicators 86
Refractive index 19
Replacement time (*see also* Residence time) 30
Residence time 26
  distribution 26, 27, 37
Resistance to gas transfer 89
Resistance to stress 186
Respiration rate 83, 108–110, 224
Respiratory enzymes 110
Respiring film thickness 223
Respirometry 81
Resting cells (*see* Non-growing cells)
Ribonucleic acid (*see* RNA)
Rich medium 118, 133
RNA 123
  turnover 70
Roller fermenter 94
Roller tubes 105

Saturation constant (K_s) 8, 10–12, 22, 64
  measurement 10
  of polymer 11
Secondary metabolite 54, 157
Seitz filter 135
Selection of organisms 45
Selenium 127
Sewage purification 50
Shake-flask 2, 94, 102–105
Shear 94
Shift in growth rate (*see* Specific growth rate)
Silicone 103

Single point estimate of growth 23
Sodium ion requirement 123
Sparger 95, 98–100
Specific growth rate (*see also* Colony growth) 4, 30
  apparent 68, 194
  effect of maintenance energy on 70–71
  effect on metabolism 74
  effect of pH on 144
  intracellular control of 44
  maximum 33, 54
  minimum 188–189
  oscillations 124
  shift down 53
  shift up 53, 218
  total 68
Specific maintenance rate 68
Specific metabolic rate (*see* Metabolic quotient)
Spores 187, 189, 195
Sporulation kinetics 189–193
Staining cells 21
Stationary liquid film theory 88
Stationary phase 23, 54, 186–188
Stimulants 169
Stirred fermenter 94, 102
Stirrer rate 98
Storage product 15, 75
Stromatic mycelium 230
Submerged film of biomass (*see* Film of biomass)
Substrate-accelerated death 187
Substrate inhibition (*see* Inhibition)
Substrate-limited growth 41, 44, 136, 231, 237
  in mixed cultures 199–210
Sulphite oxidation 90
Sulphonamides 174
Sulphur nutrition 124–125
Surface active agents (*see* Surfactants)
Surfactants 80, 100
Symbiosis 199
Synchrony of cell division 24
Synthetic medium 117

Teflon 40
Teichoic acid 123
Teichuronic acid 123
Temperature effects 137–142, 197, 250
Temperature characteristic (*see* Activation energy)
Test tube aeration 105
Thermophile 140
Tonicity 147, 149–155
  of cell contents 150
  measurement 149

Tower fermenter 48
Toxic effects (*see* Inhibition)
Trace elements 117, 125–128
   removal from media 128
Transient state 40, 162, 164, 213, 218
Trickling filter 224
Turbidity measurement 43
Turbidostat 42–45
Turbulence 95–96
Turnover rate
   of biomass 68
   of cell walls 70
   of protein 70
   of RNA 70

Urinary infection 211

Vaned disc impeller 97
Vegetative cell 57

Viability 58, 187
   index 59–61
Viable count 21
Viable fraction 58–62
Vibrating agitator 94
Viscosity 98, 101
Vital stain 21
Vitamers 121
Vitamins 1, 117, 121–123
Vortex aeration 95

Waldhof fermenter 96
Wall growth 39–40
Water activity 147–155
   measurement 149
Winkler method 84

Y_{ATP} (*see* ATP yield)

Zero growth rate effects 54, 186–193